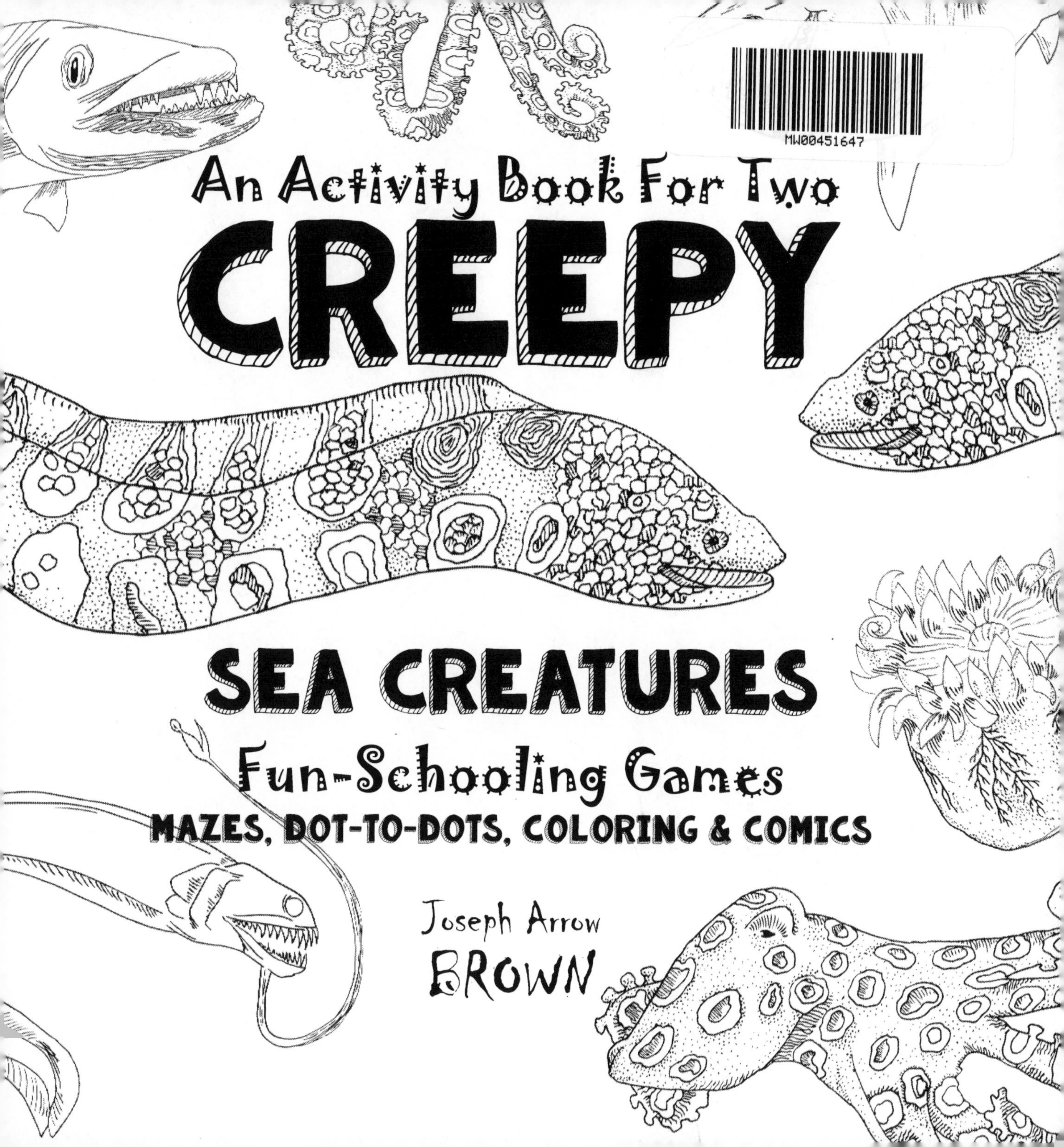

An Activity Book For Two

CREEPY SEA CREATURES

Fun-Schooling Games

MAZES, DOT-TO-DOTS, COLORING & COMICS

Joseph Arrow BROWN

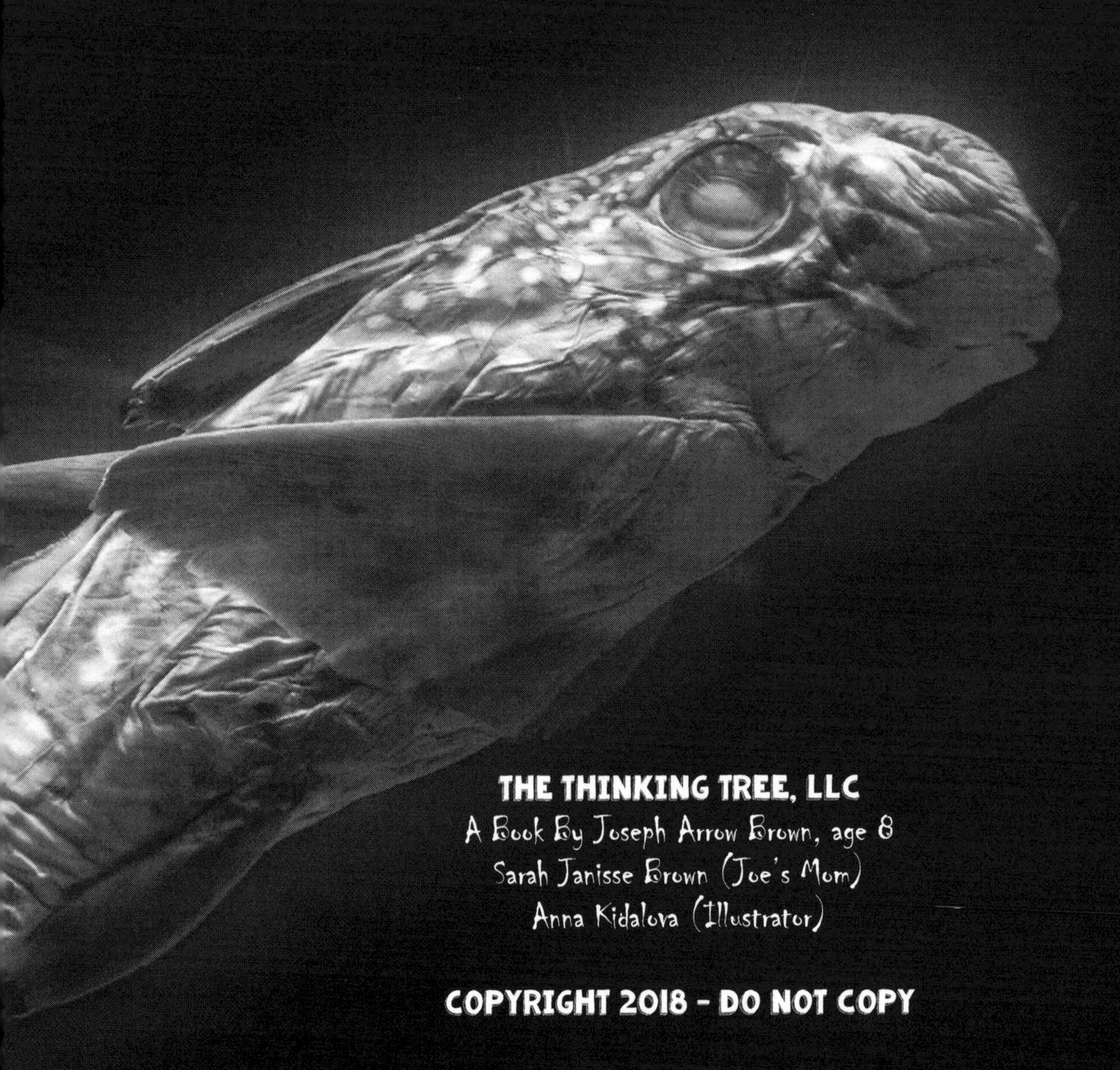

THE THINKING TREE, LLC

A Book By Joseph Arrow Brown, age 8
Sarah Janisse Brown (Joe's Mom)
Anna Kidalova (Illustrator)

FunSChoolingBooks.com

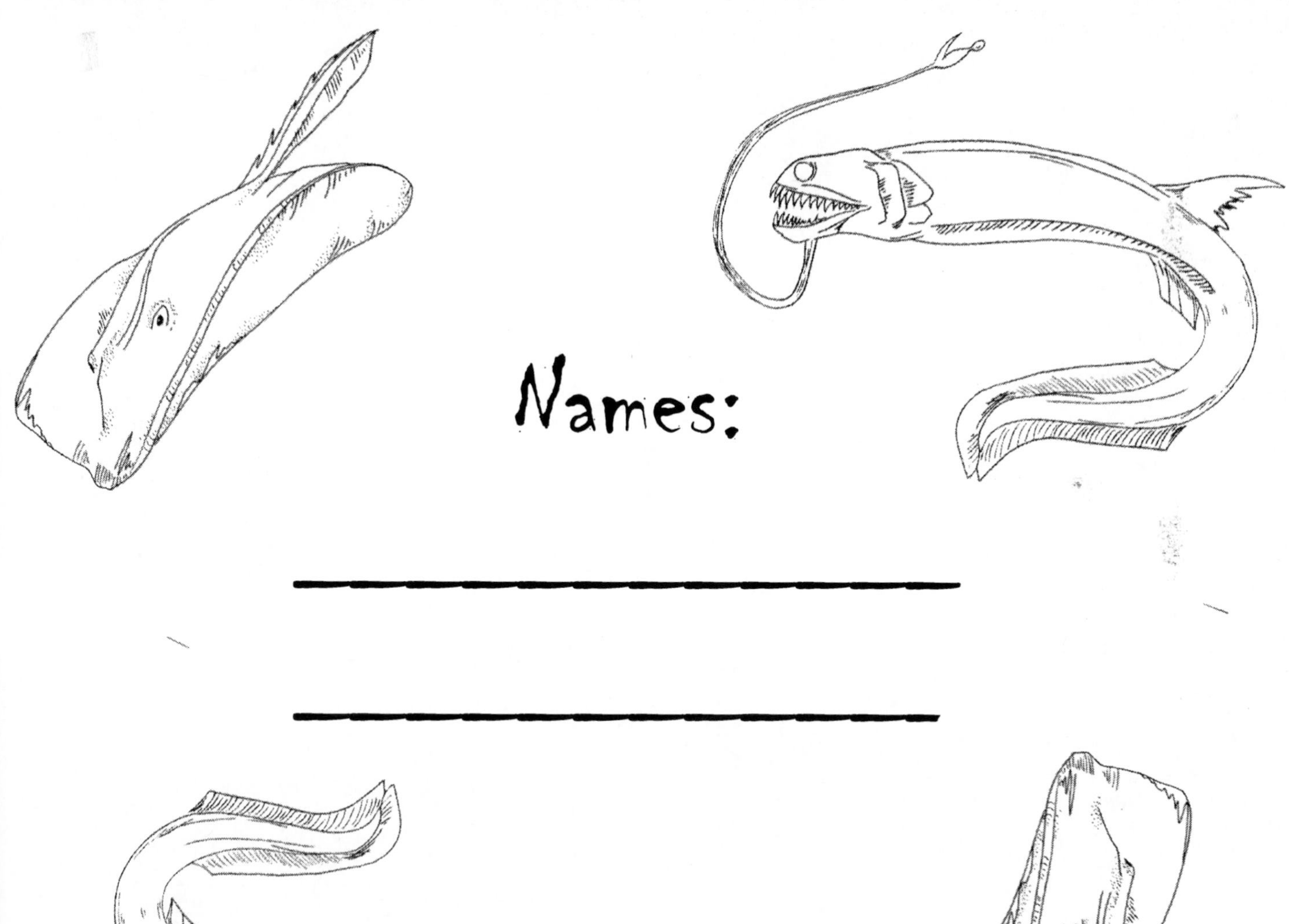

Names:

____

____

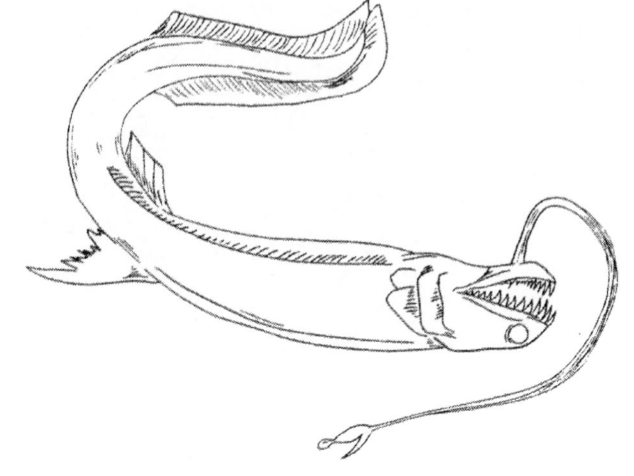

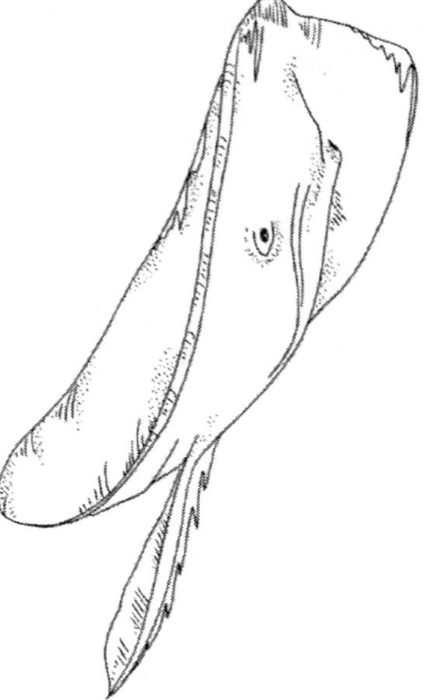

Add words to create a comic!

Race Though the Maze

Draw my Food, my Habitat & my Enemies

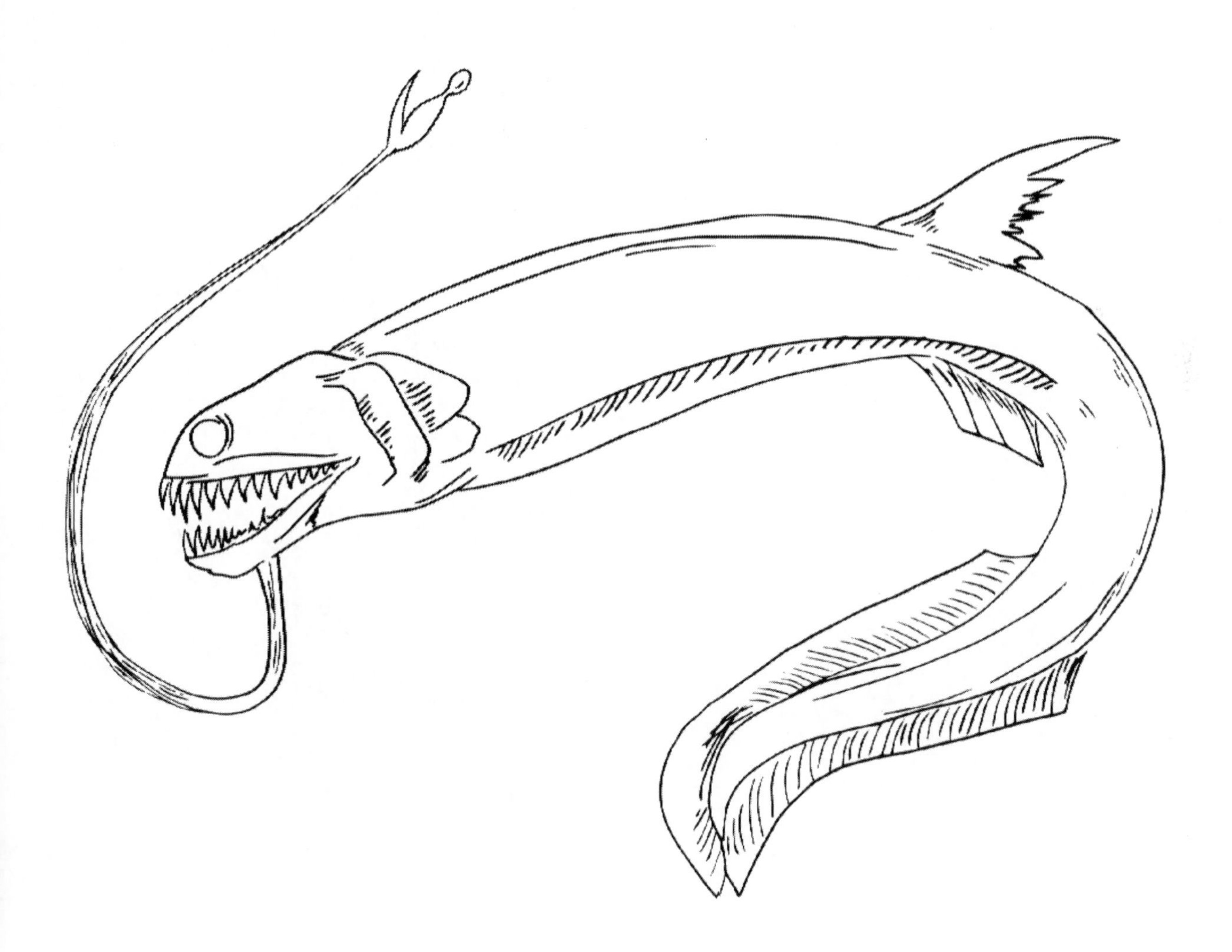

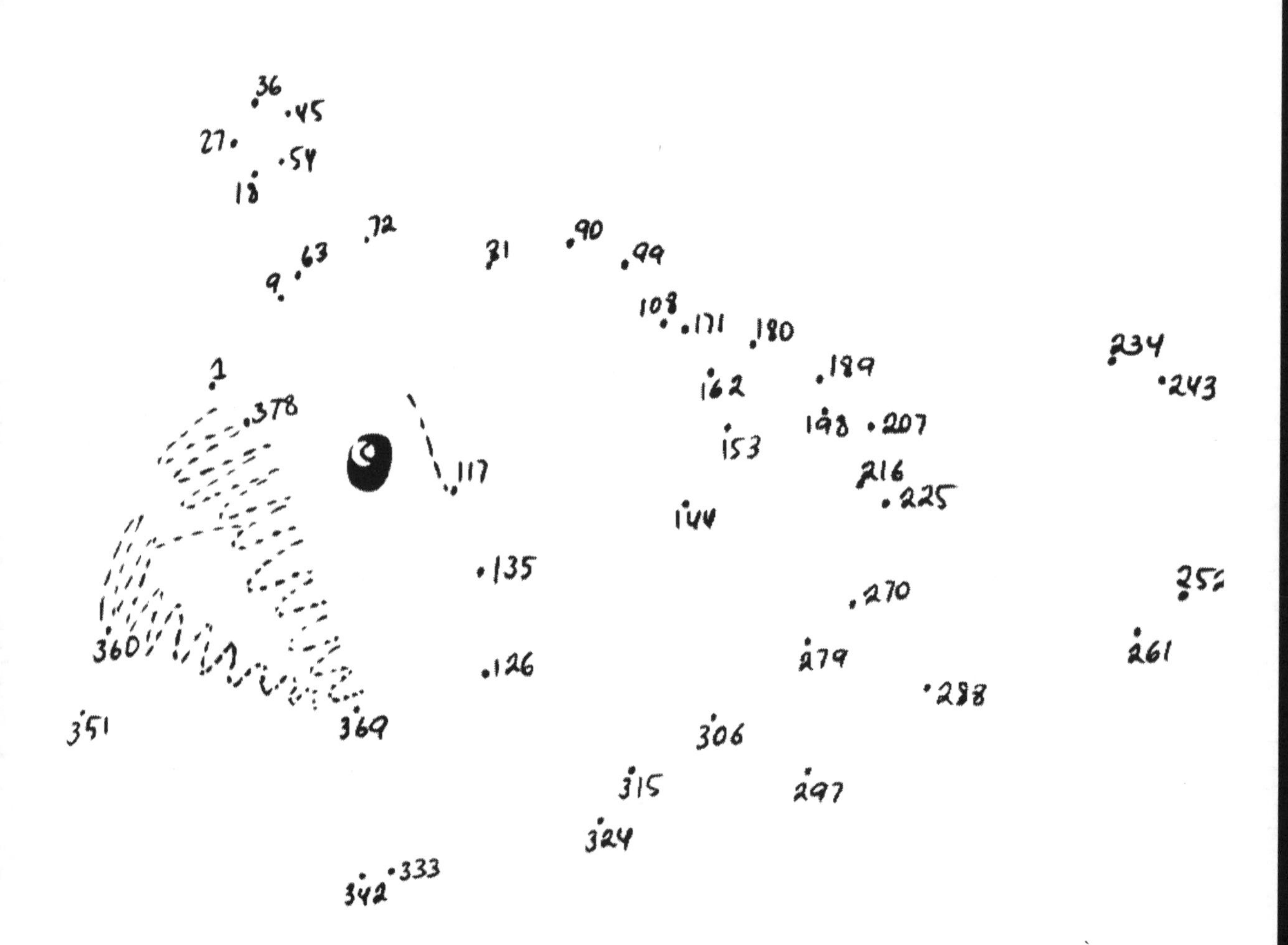

Skip-Count to see the Picture

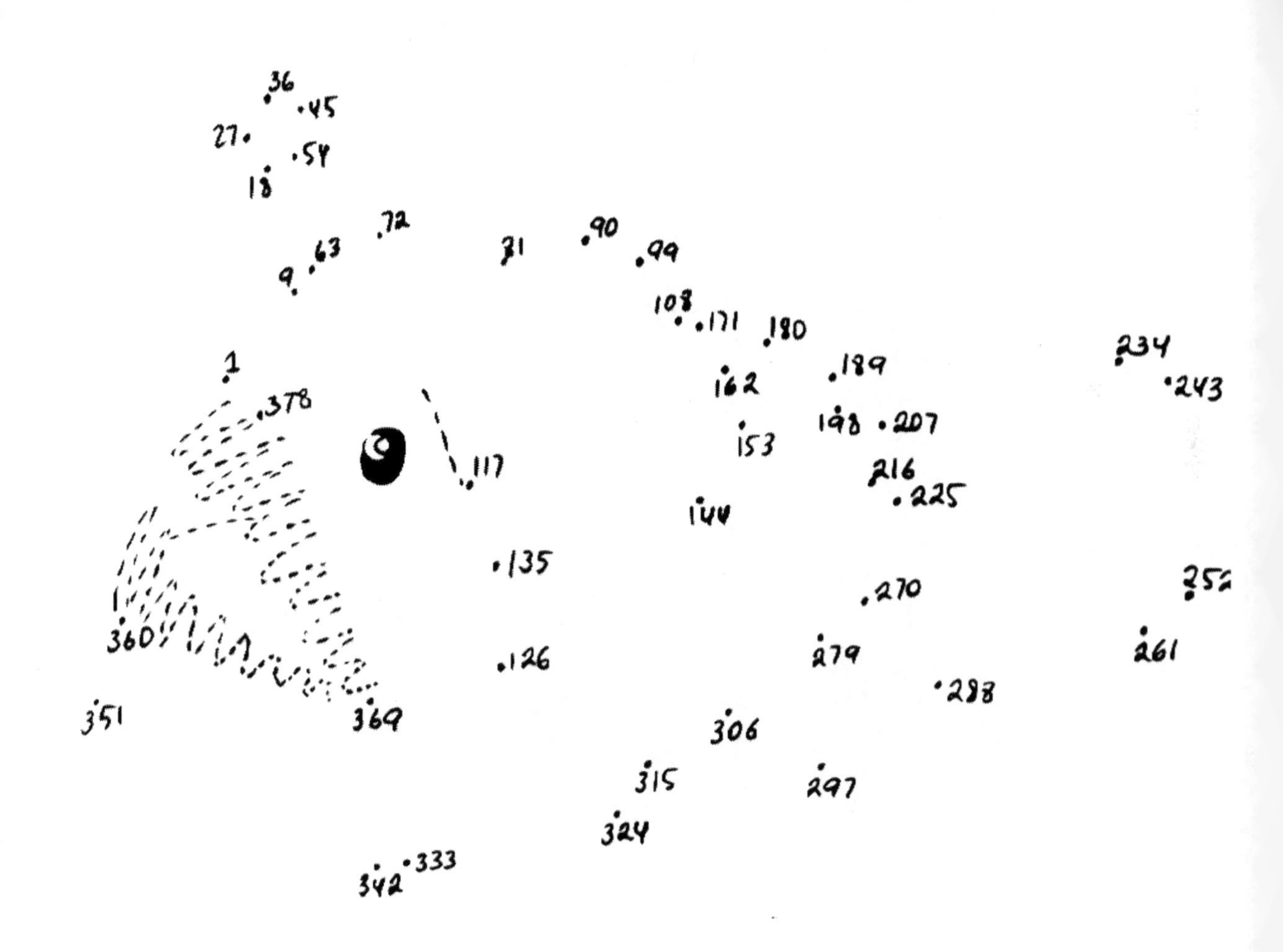
36
45
27
54
18
72
63
9
81
90
99
108
171
180
234
1
162
189
243
378
198
207
117
153
216
225
144
135
270
360
279
126
261
288
351
369
306
315
297
324
333
342

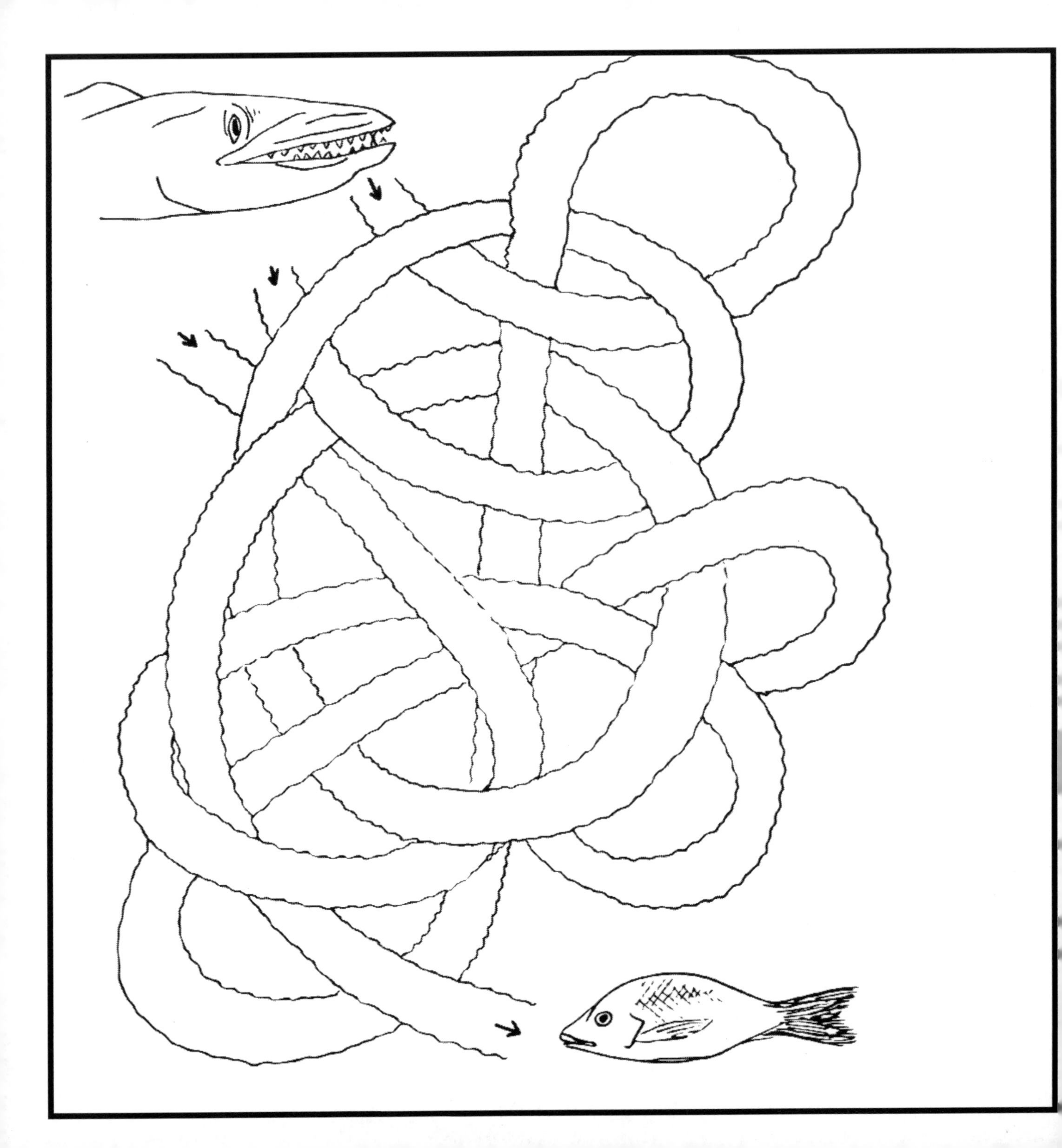

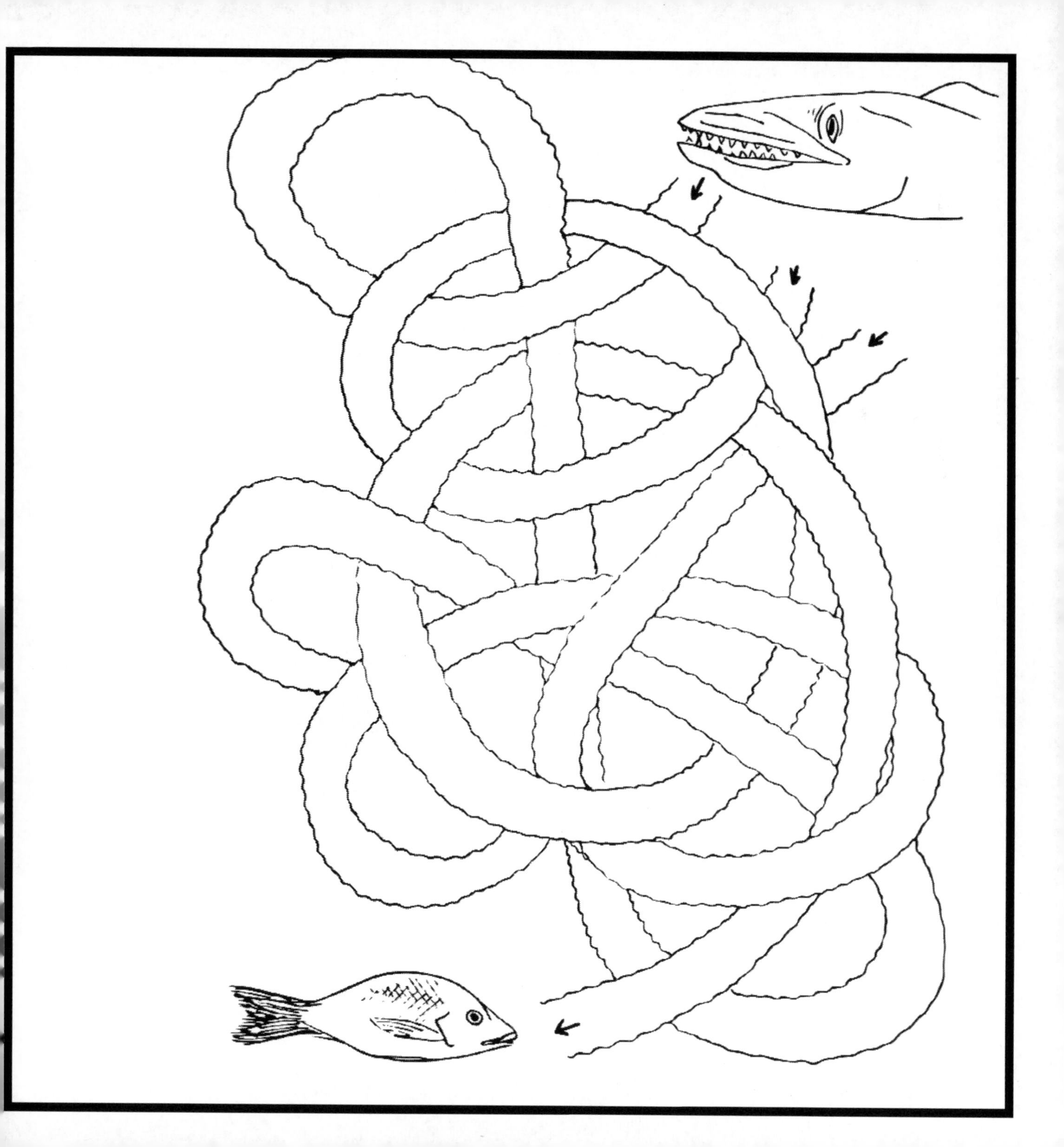

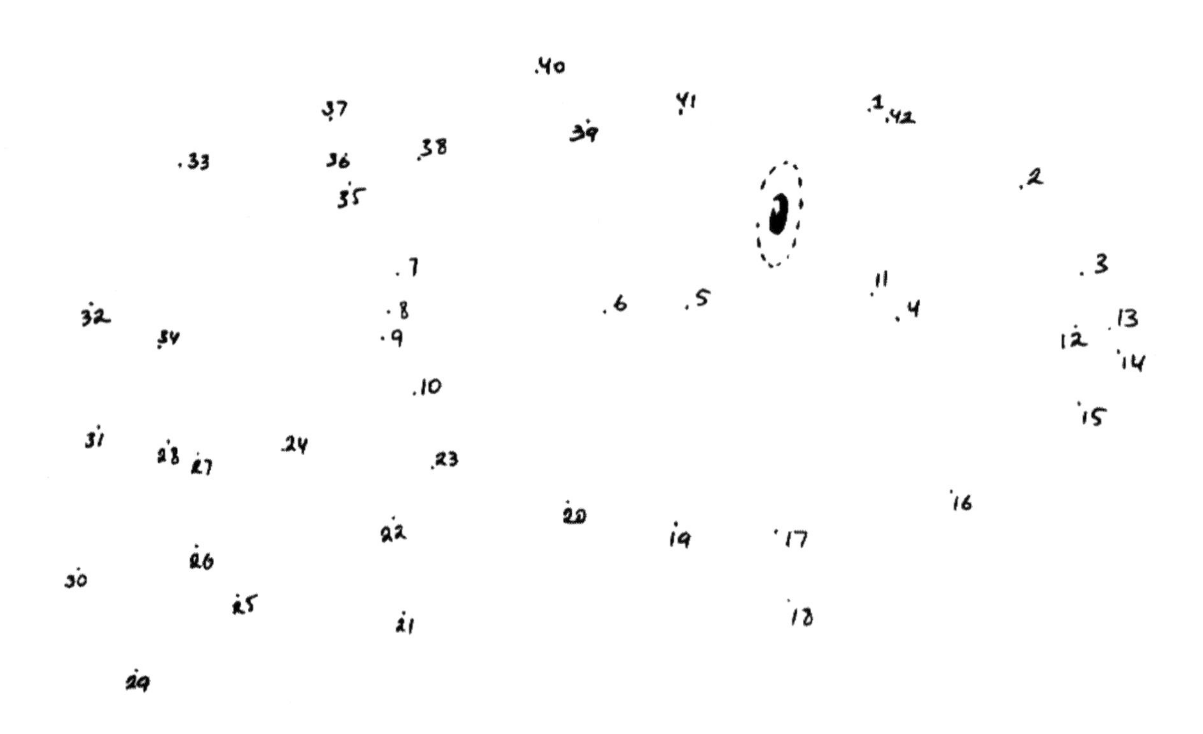

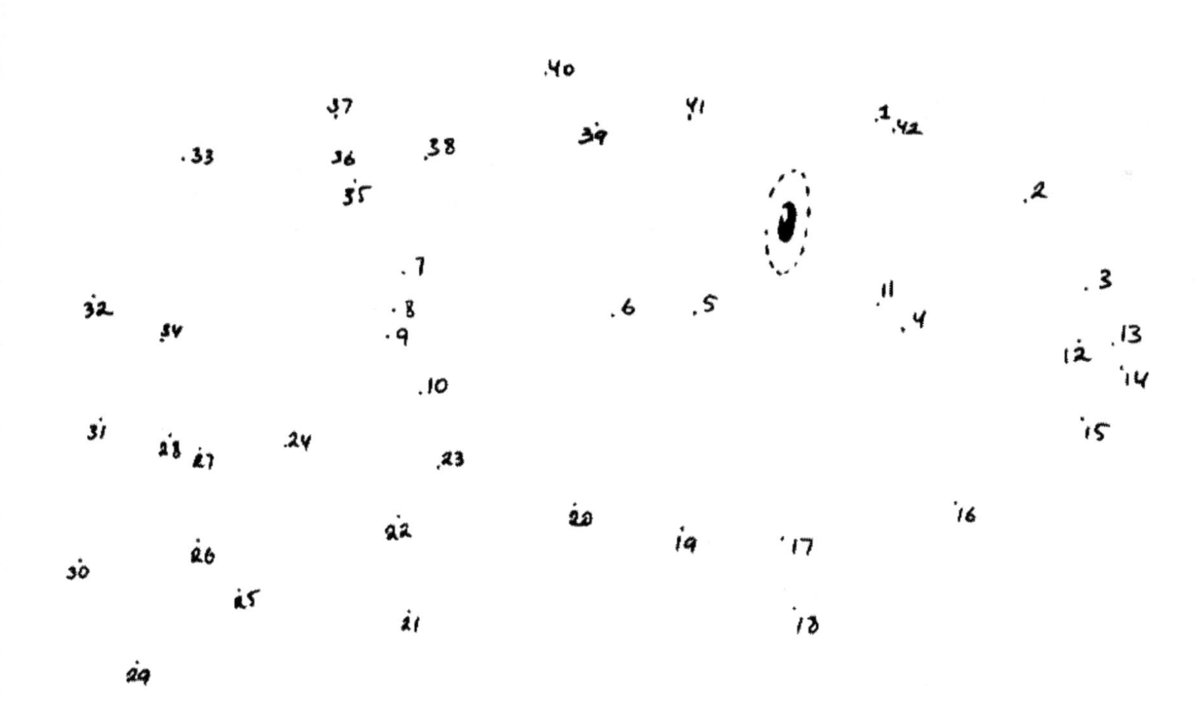

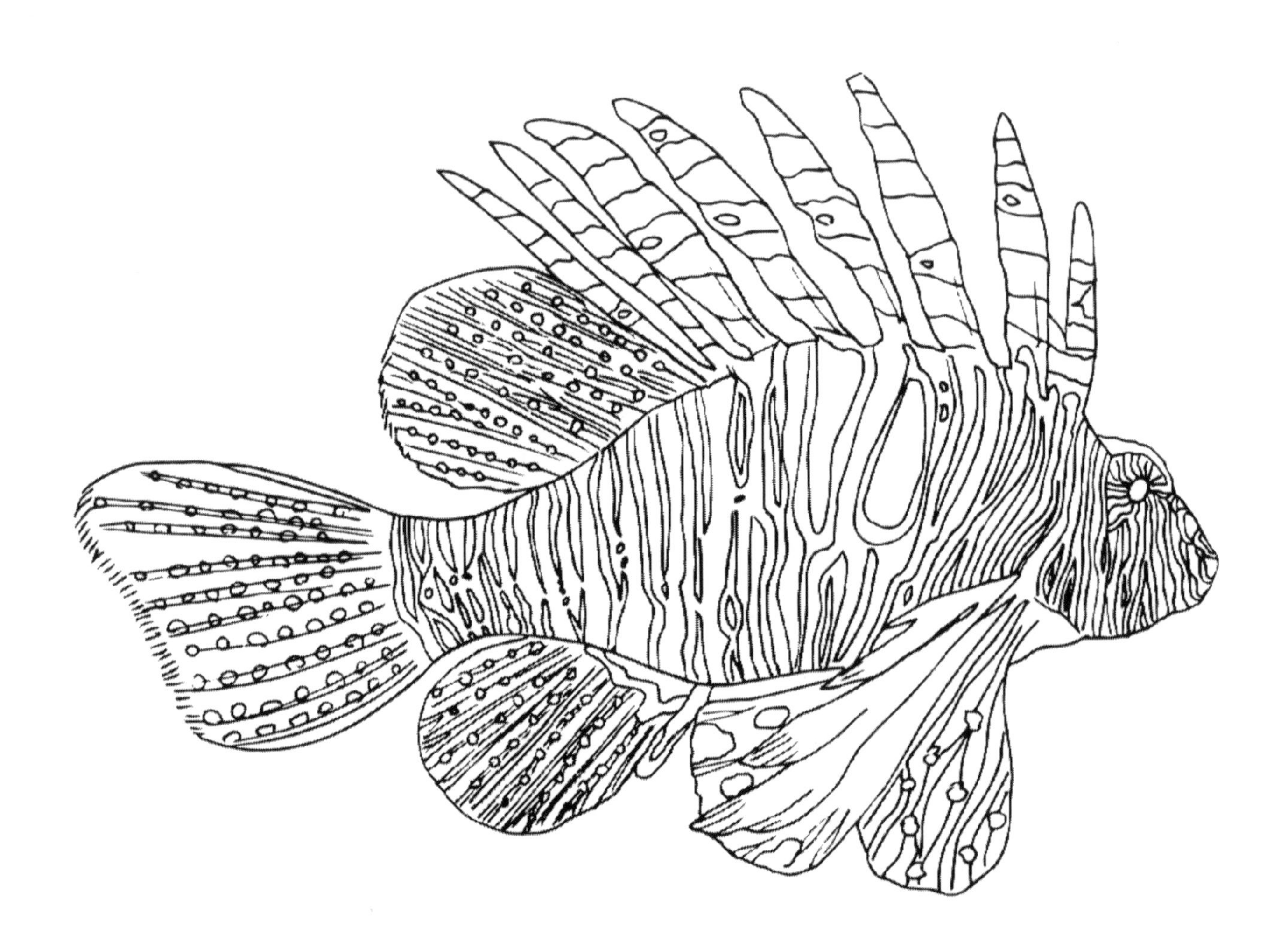

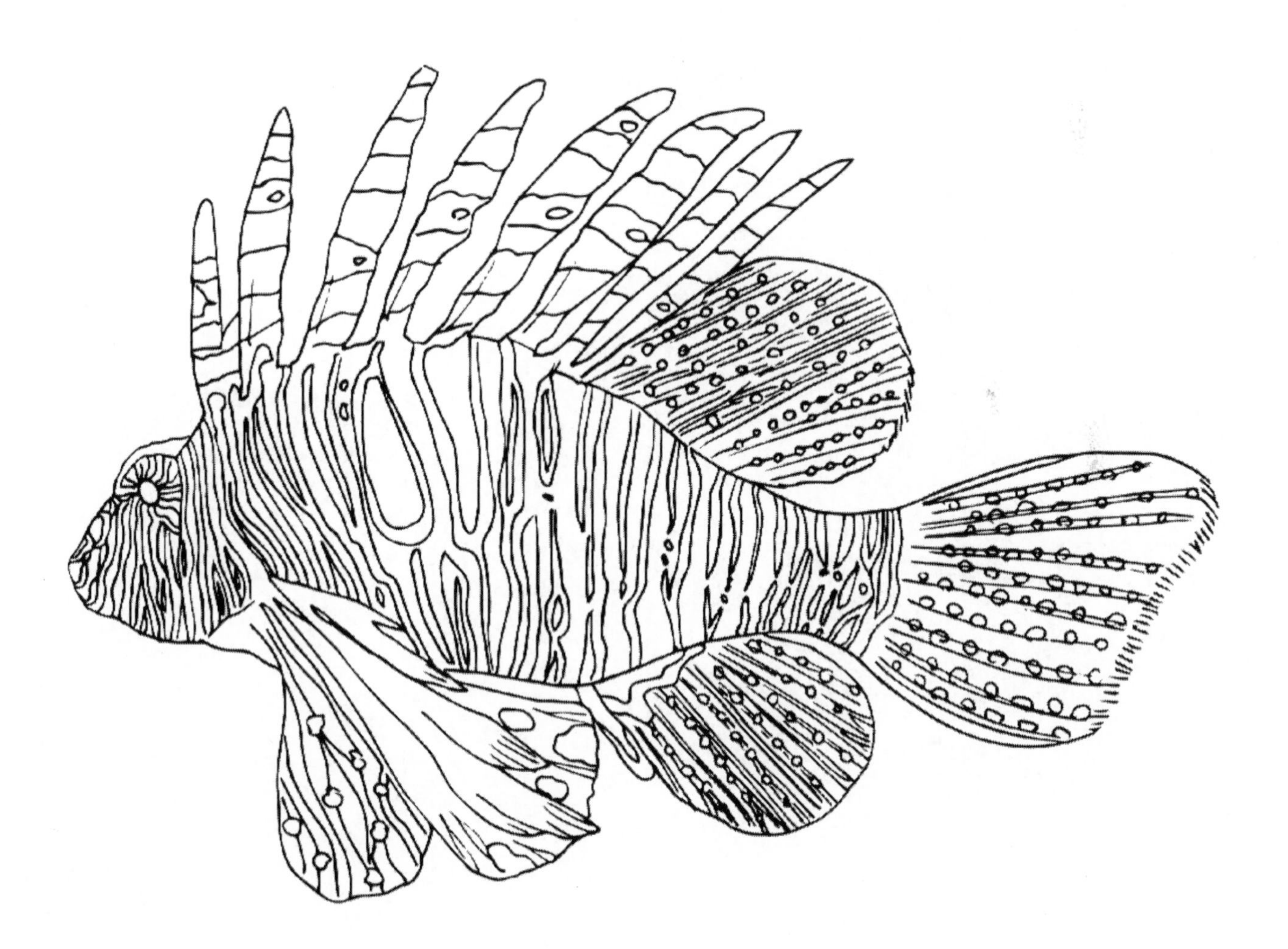

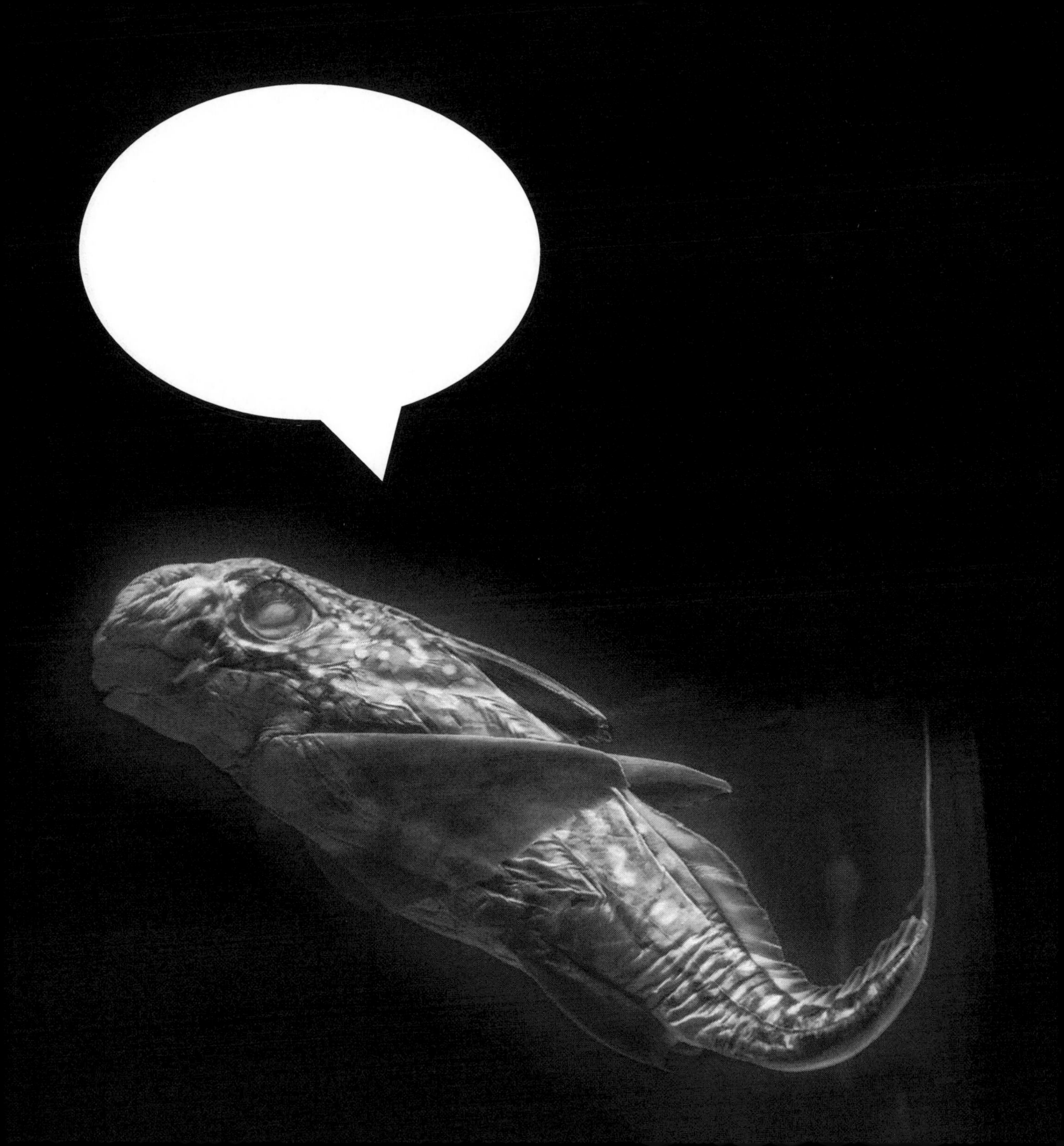

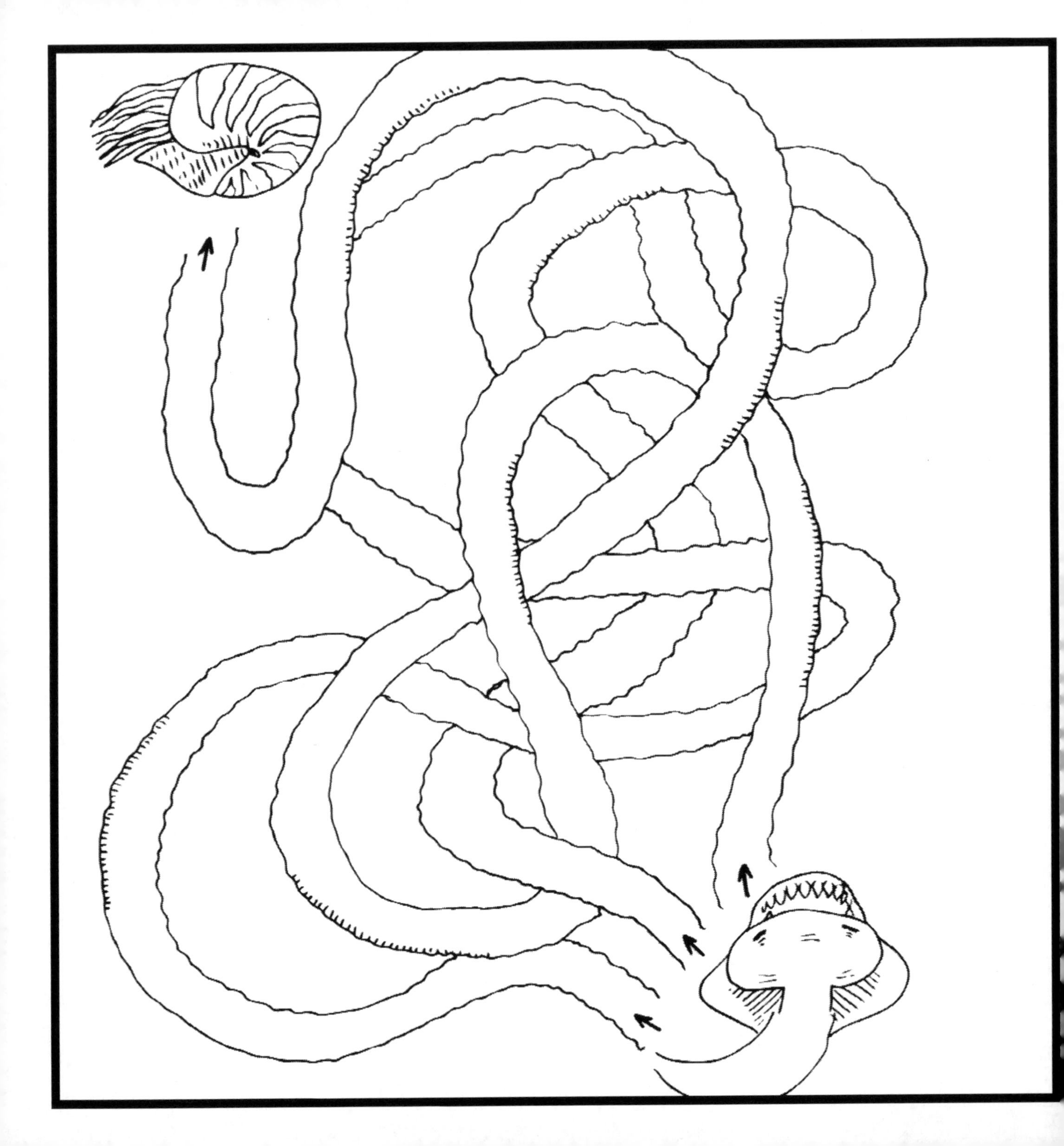

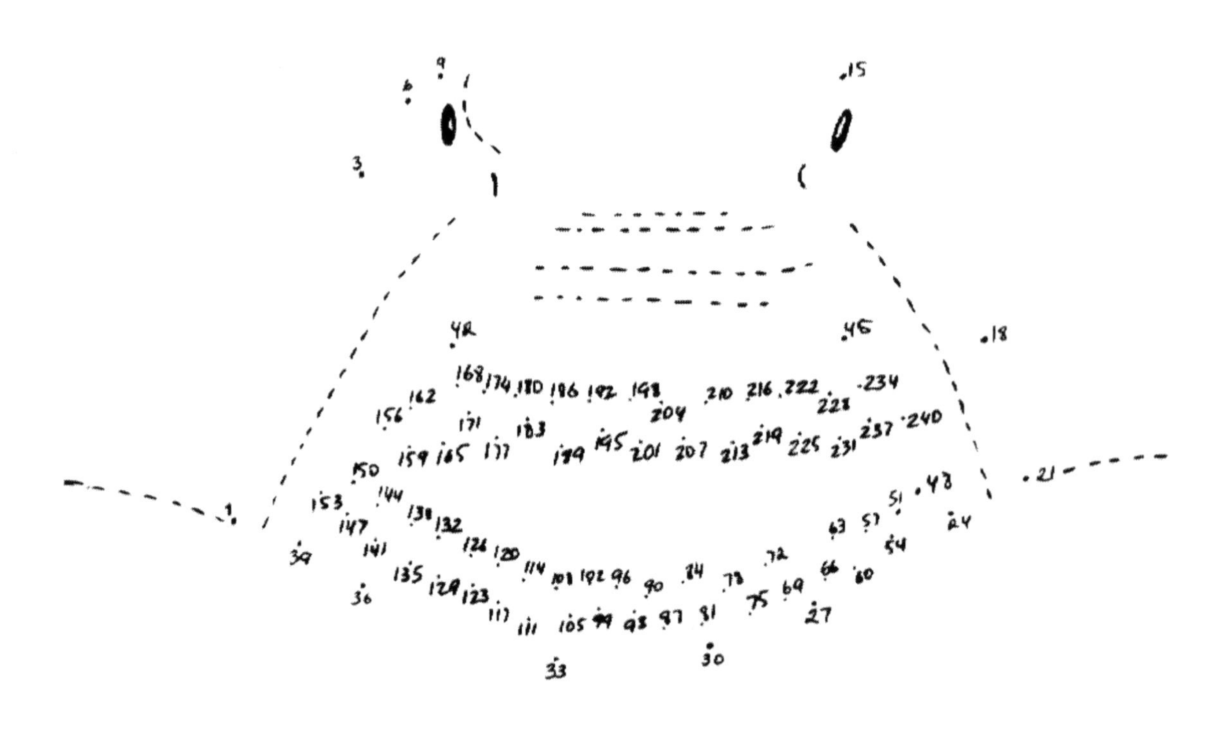

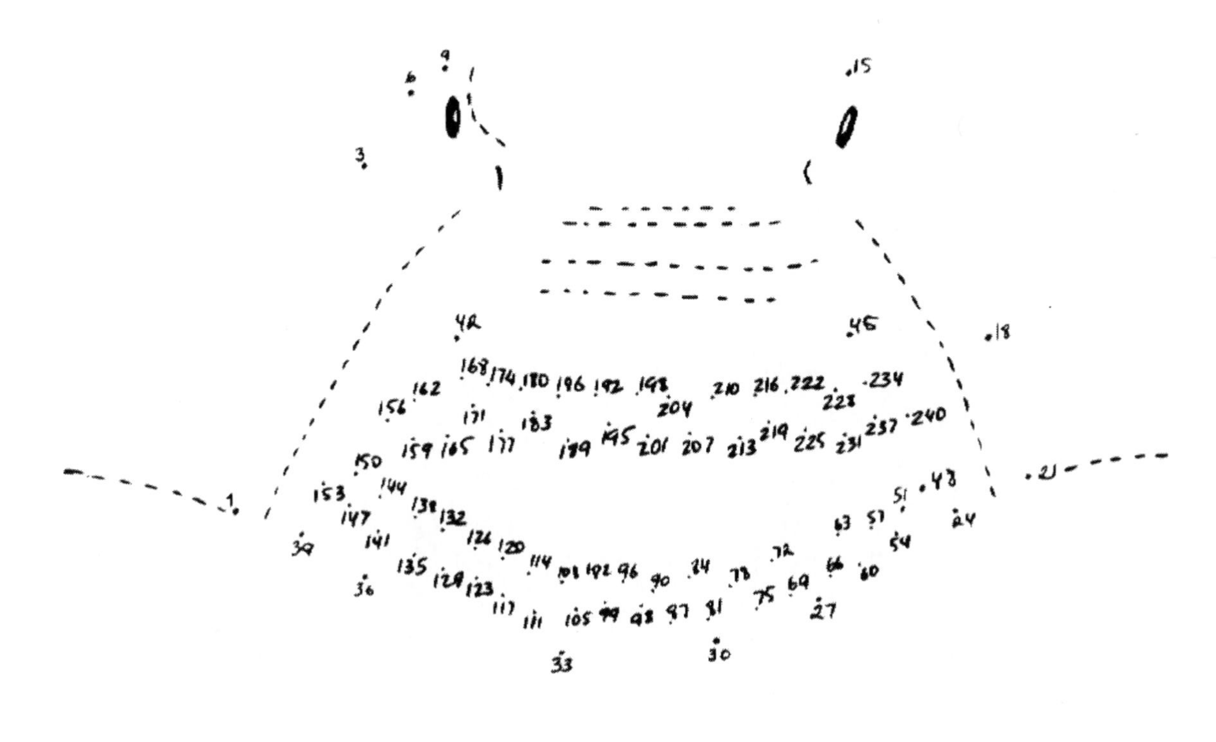

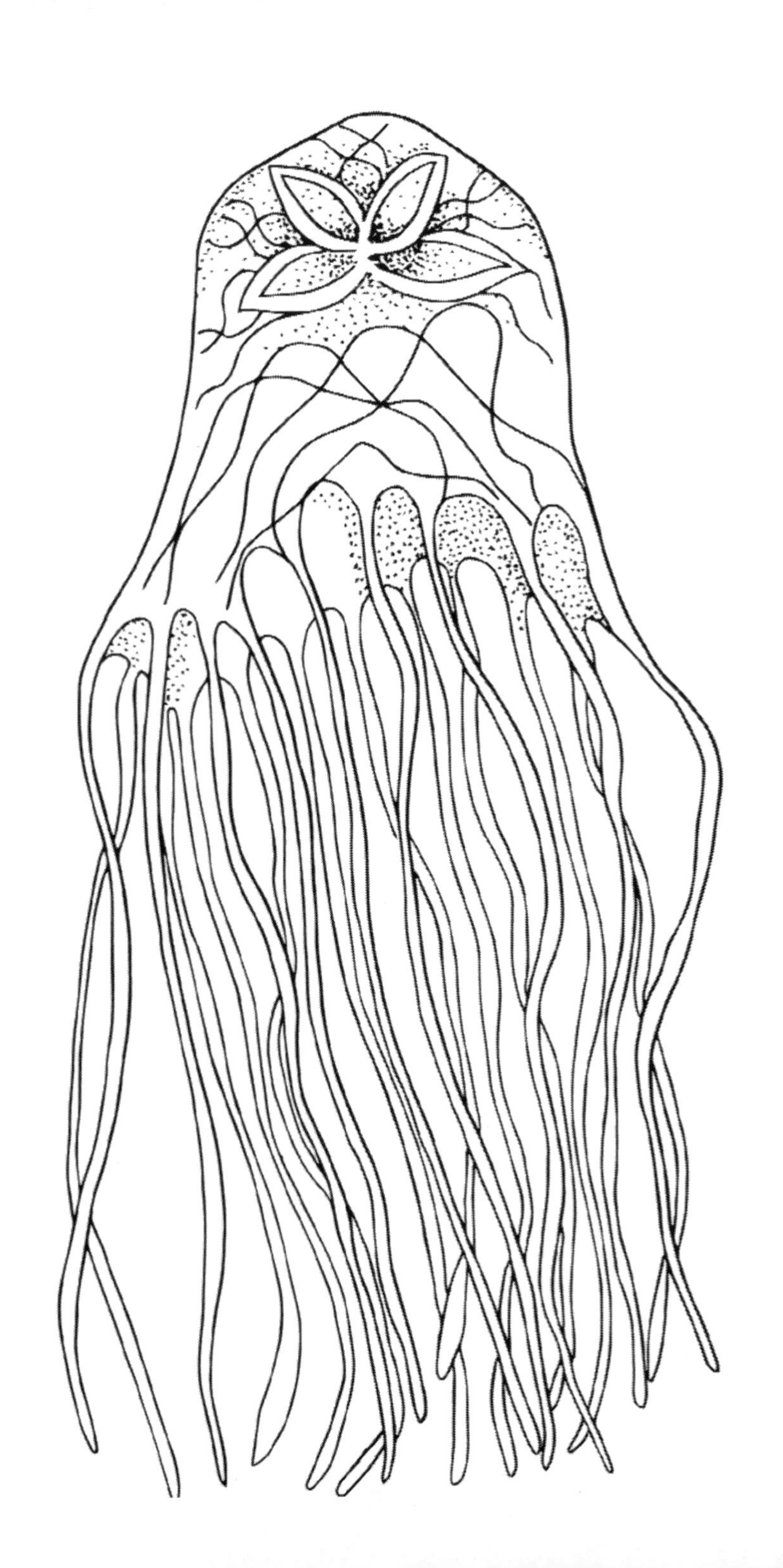

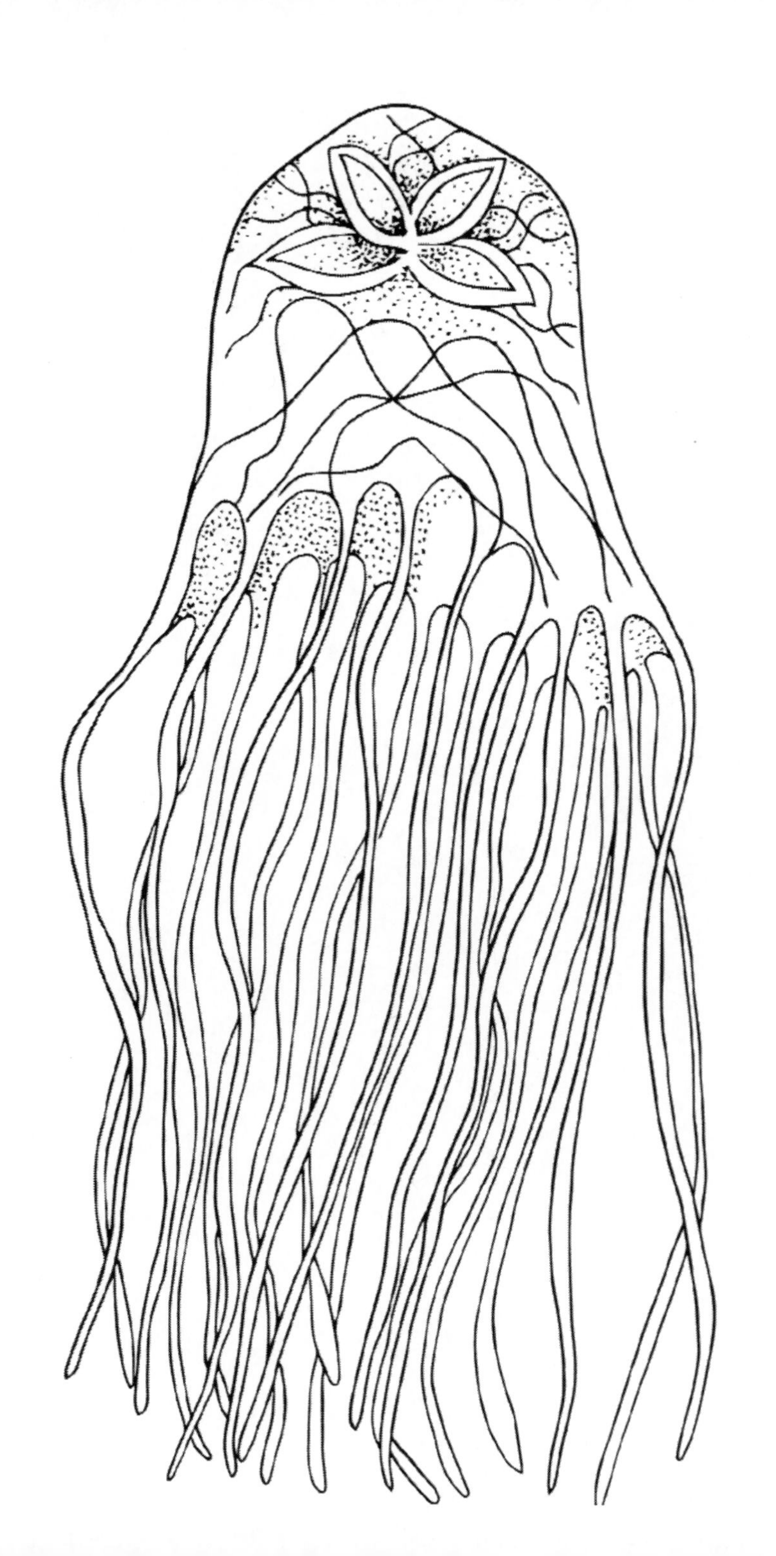

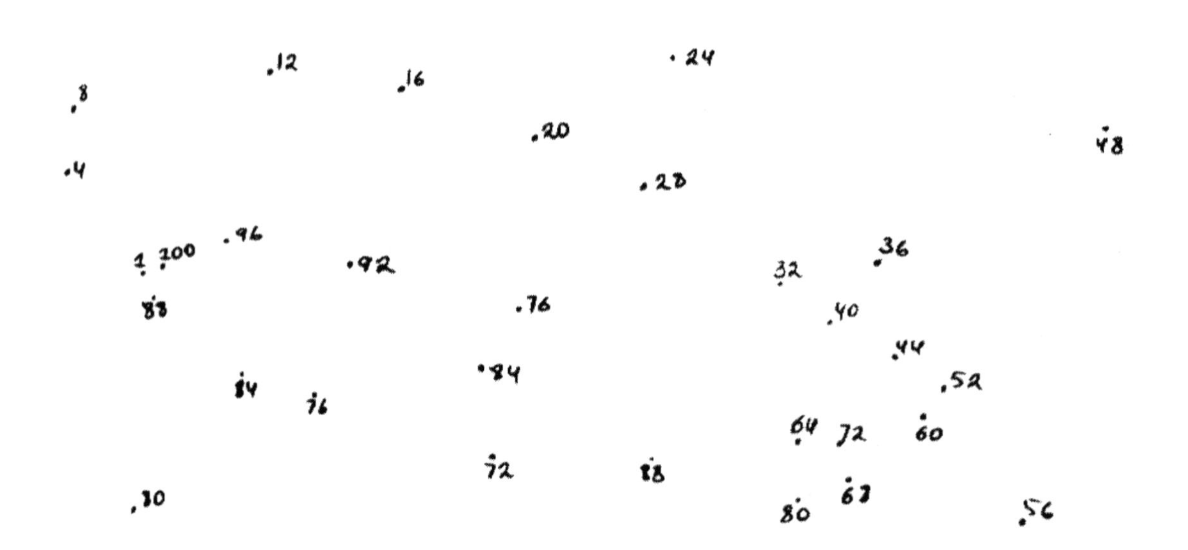

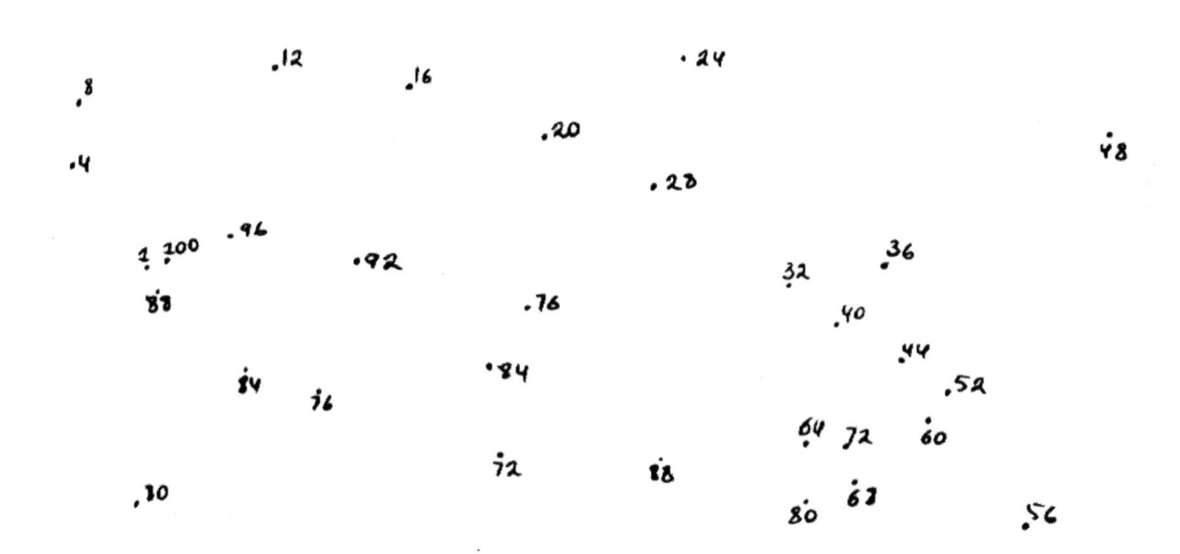

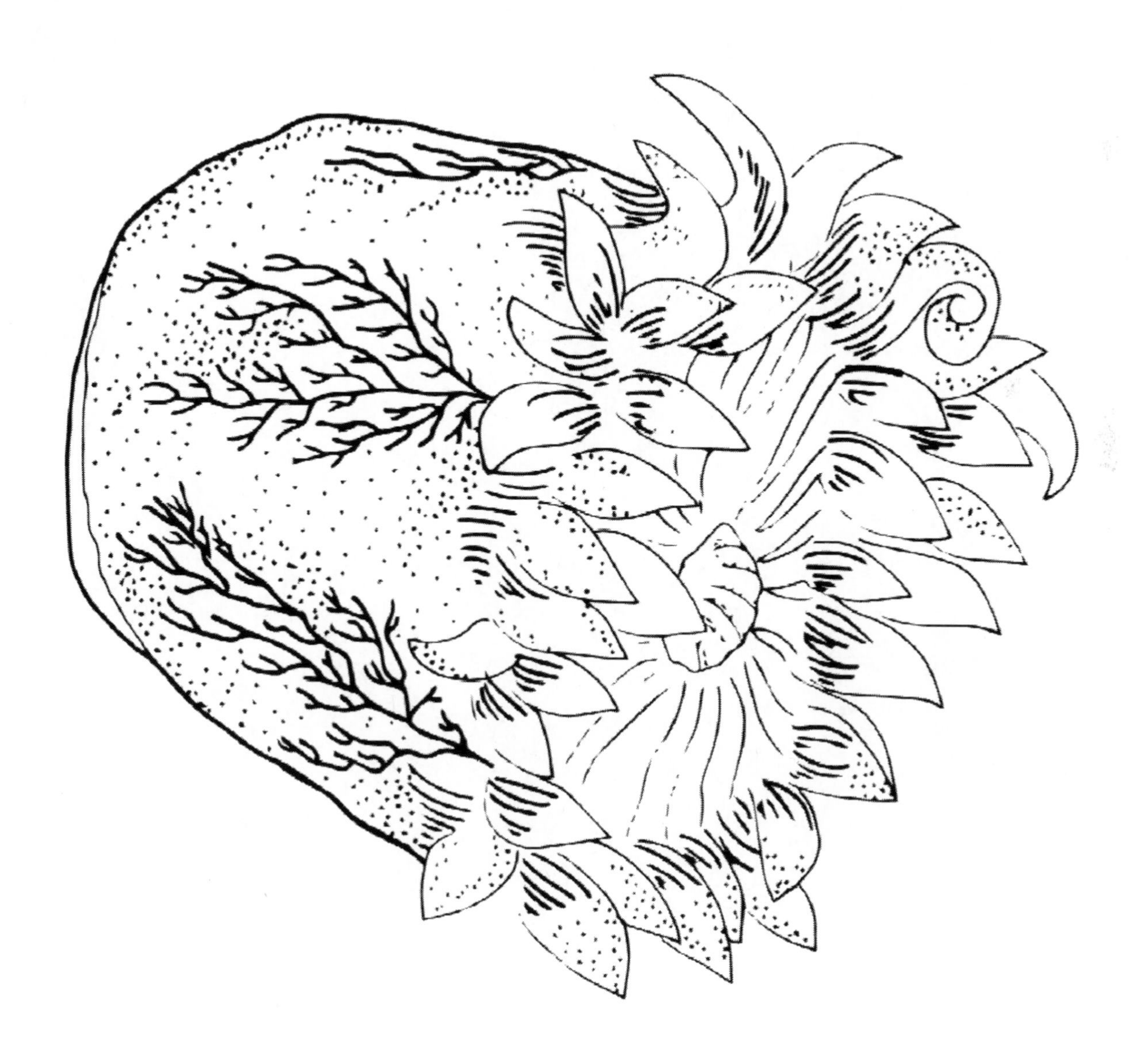

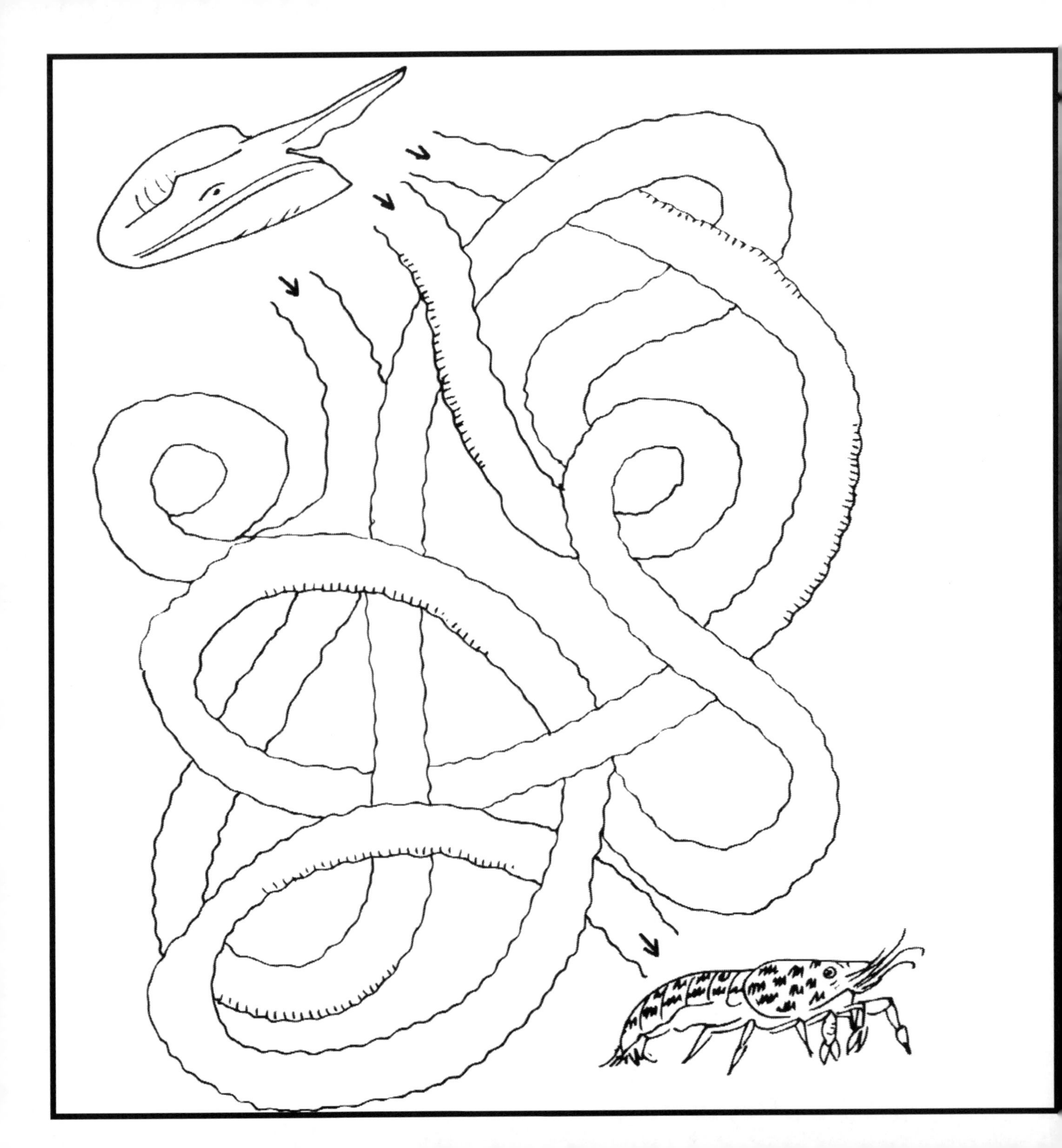

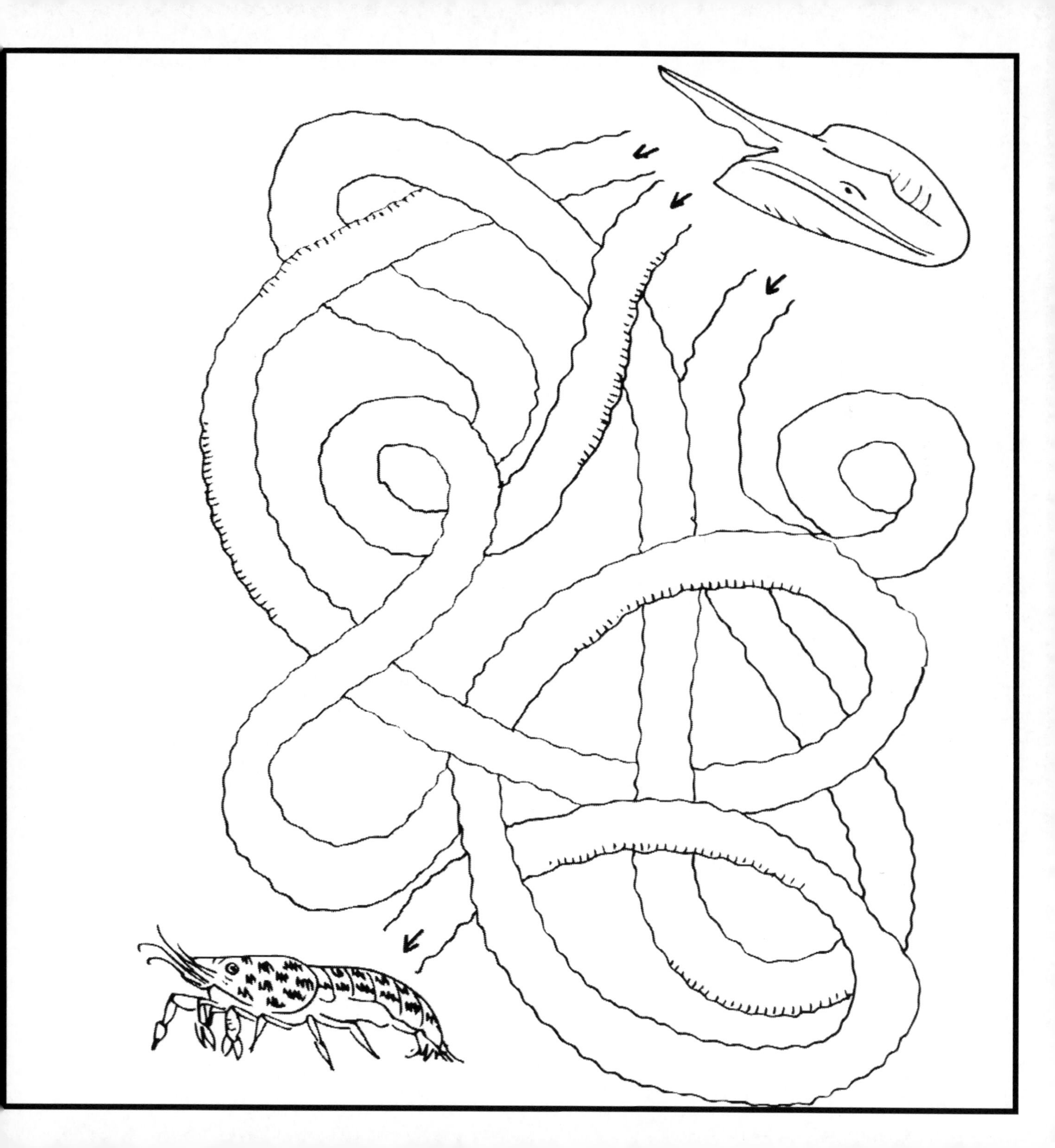

105 100

95

110 90 65

70 60

115

165 160 85 30 35

170 135 140 155 150 75 45 40 55 25

175 130 145 80 20 50 15 10 5

125 120 1

180

185 190 195

440

200

205 275 335 340 430 435

210 265 270

215 260 280 330 310 305 385 410

315 300 345 425 405

220 255 415

285 325 320 295 380 390 420 400

225 290 350 375 370 365 395

250

230 245 240 355 360

235

105 100

110 90 95 65

165 115 70 60

160 85 30 35

170 135 140 155 75 45 40 55

150 25

175 130 145 90 20 50 15 10

125 120 5

180 1

185 190 195

200 440

205 275 335 340 430 435

210 265 270

215 260 280 330 310 305 385 410

220 255 315 300 345 425 415 405

225 285 325 320 295 380 390 420 400

250 290 350 375 370 365 395

230 245 240 355 360

235

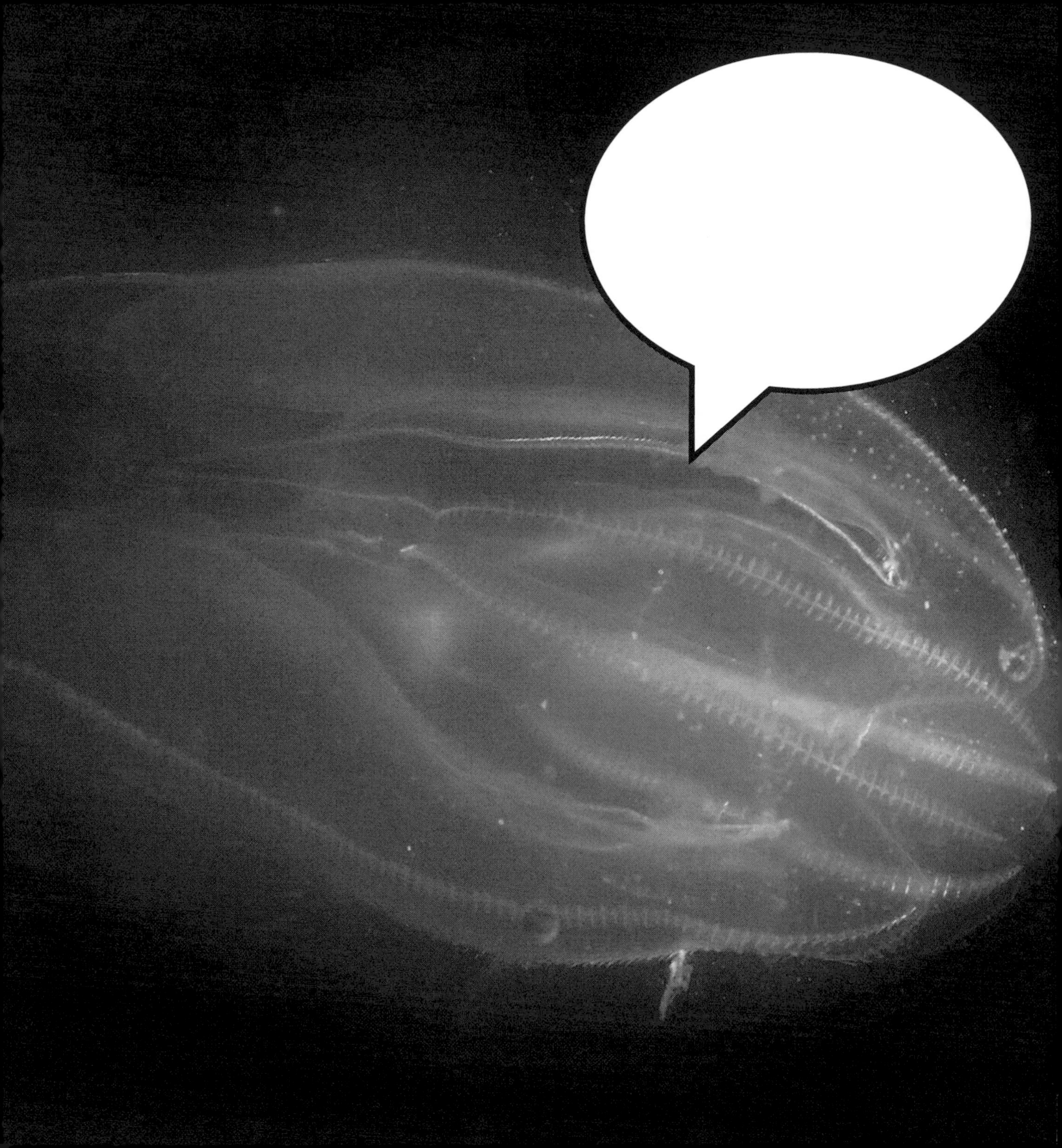

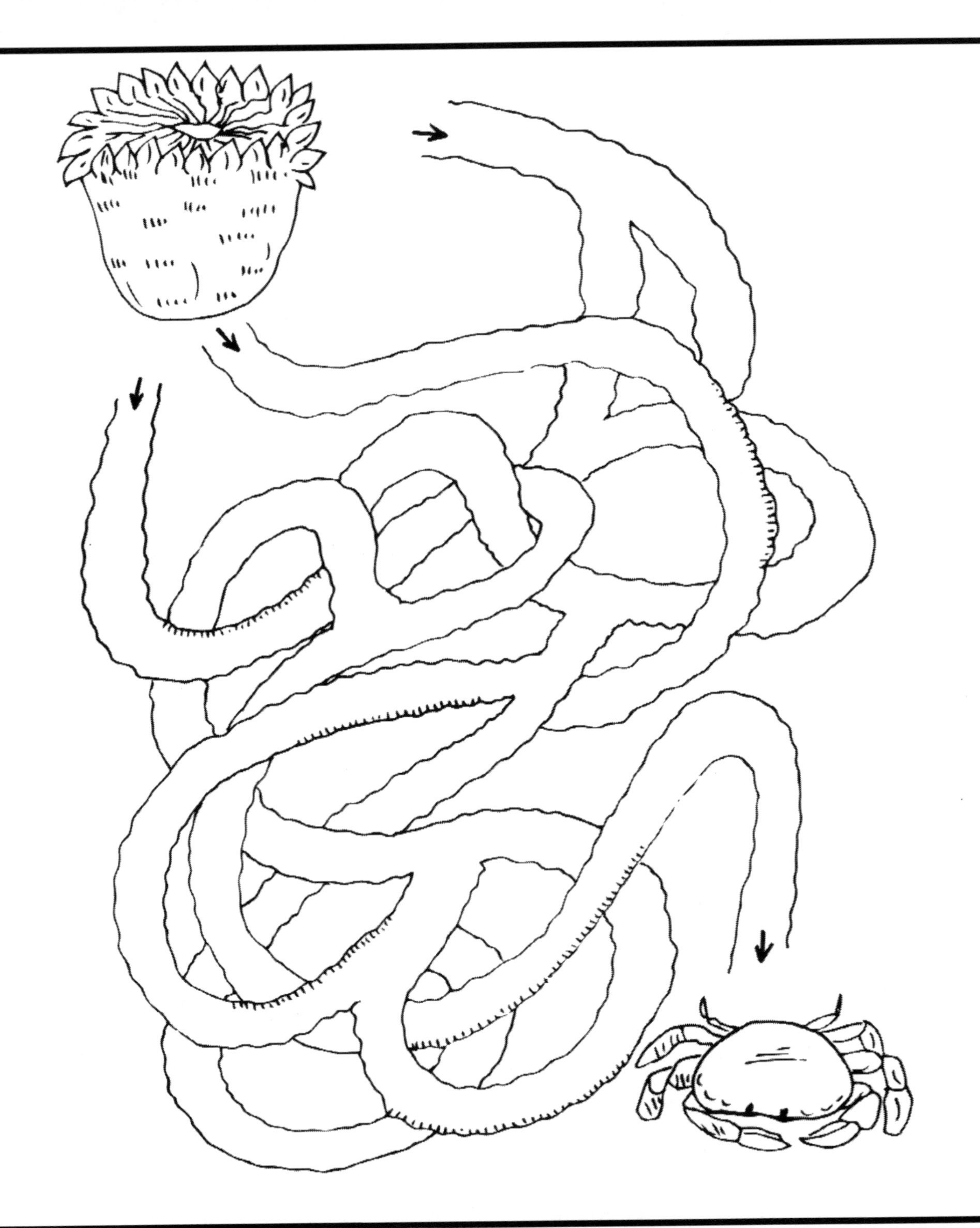

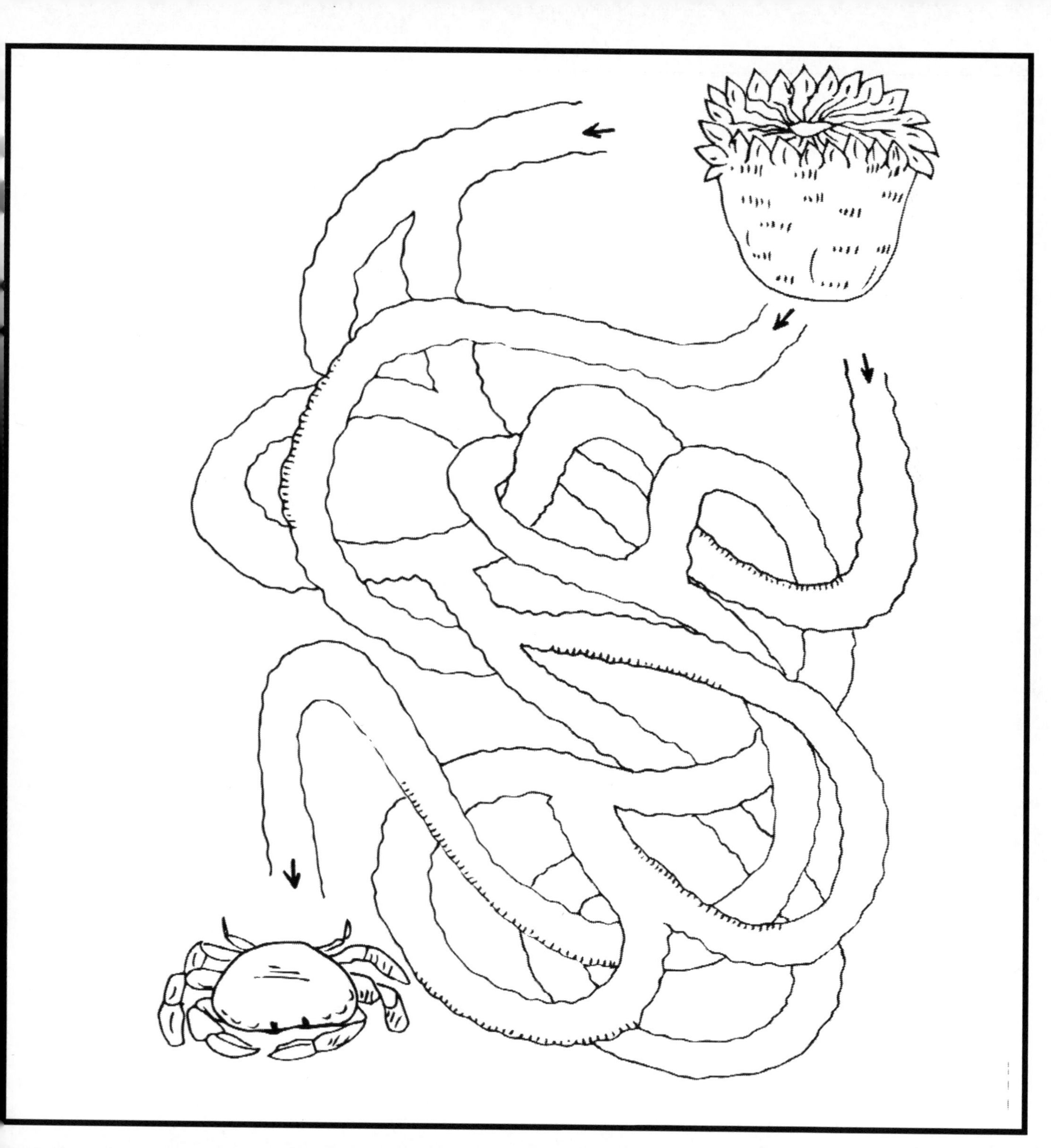

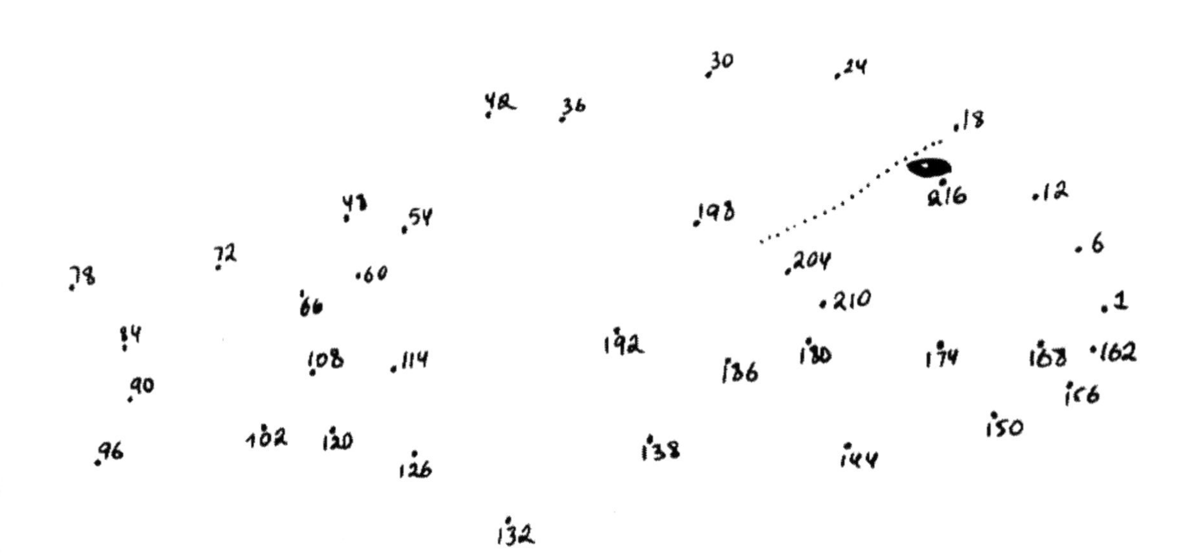
30
24
42
36
18
216
12
48
54
198
6
72
78
60
204
66
210
1
84
192
108
114
180
174
168
162
186
90
156
150
102
120
96
138
144
126
132

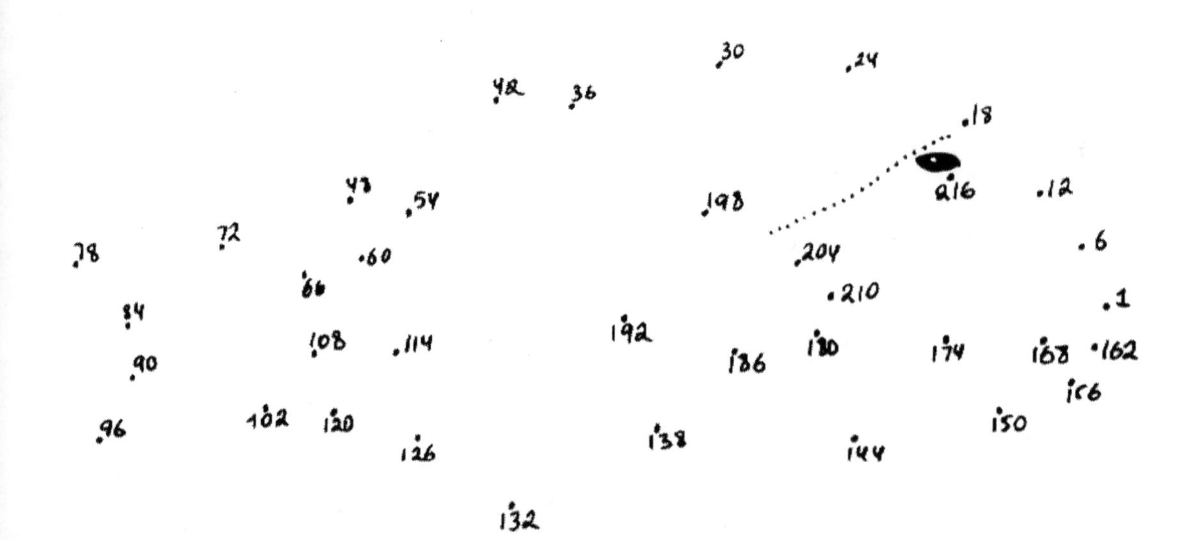

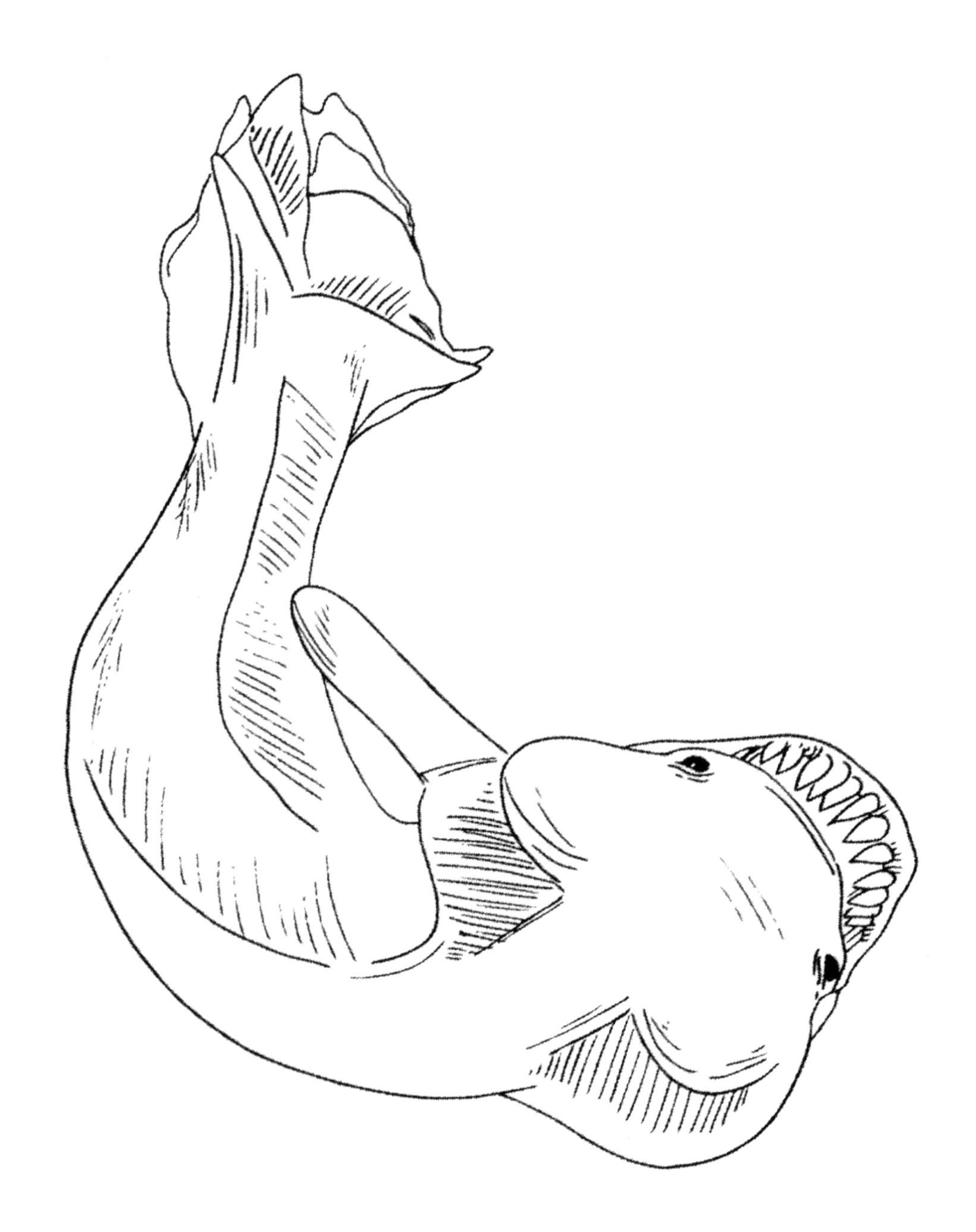

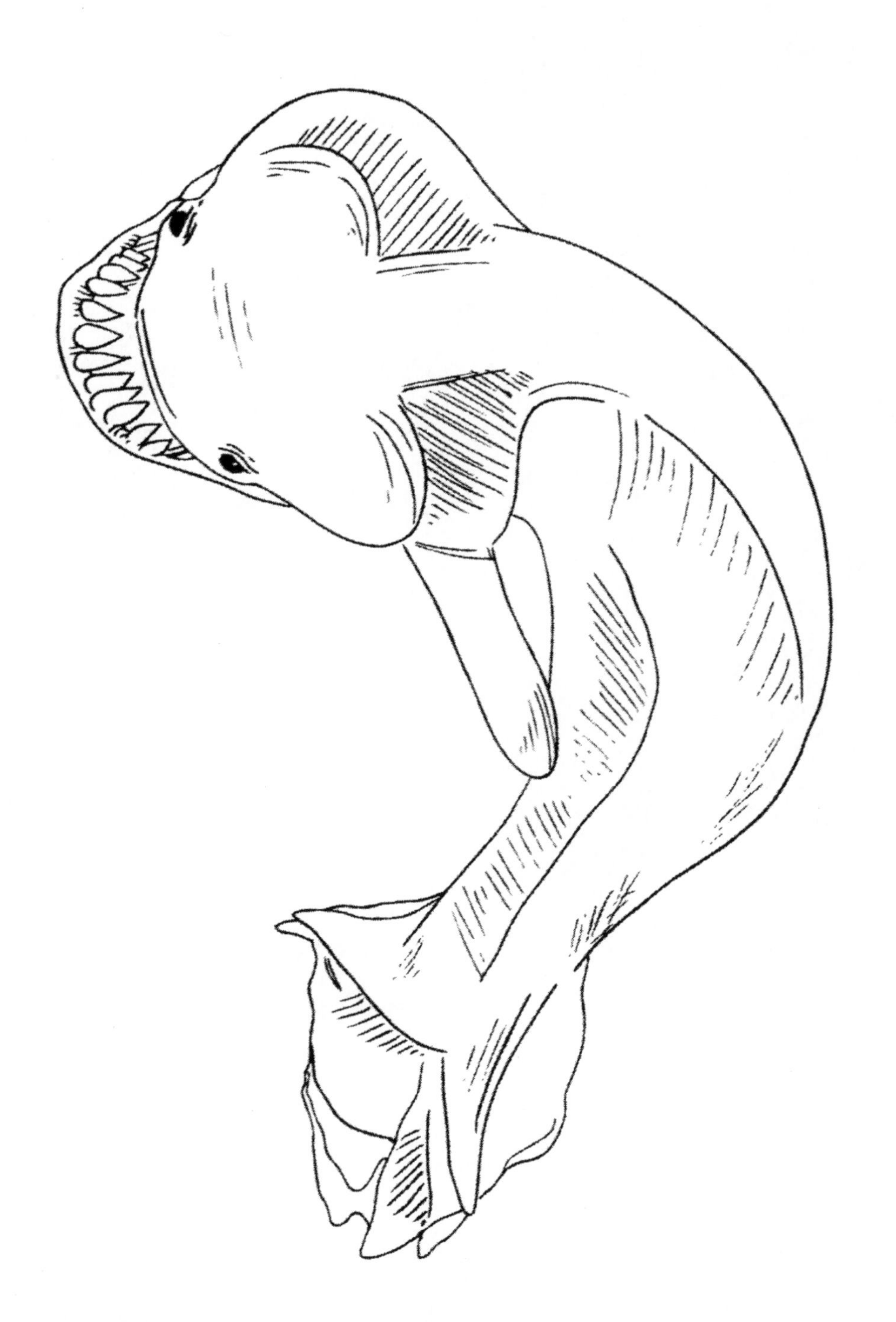

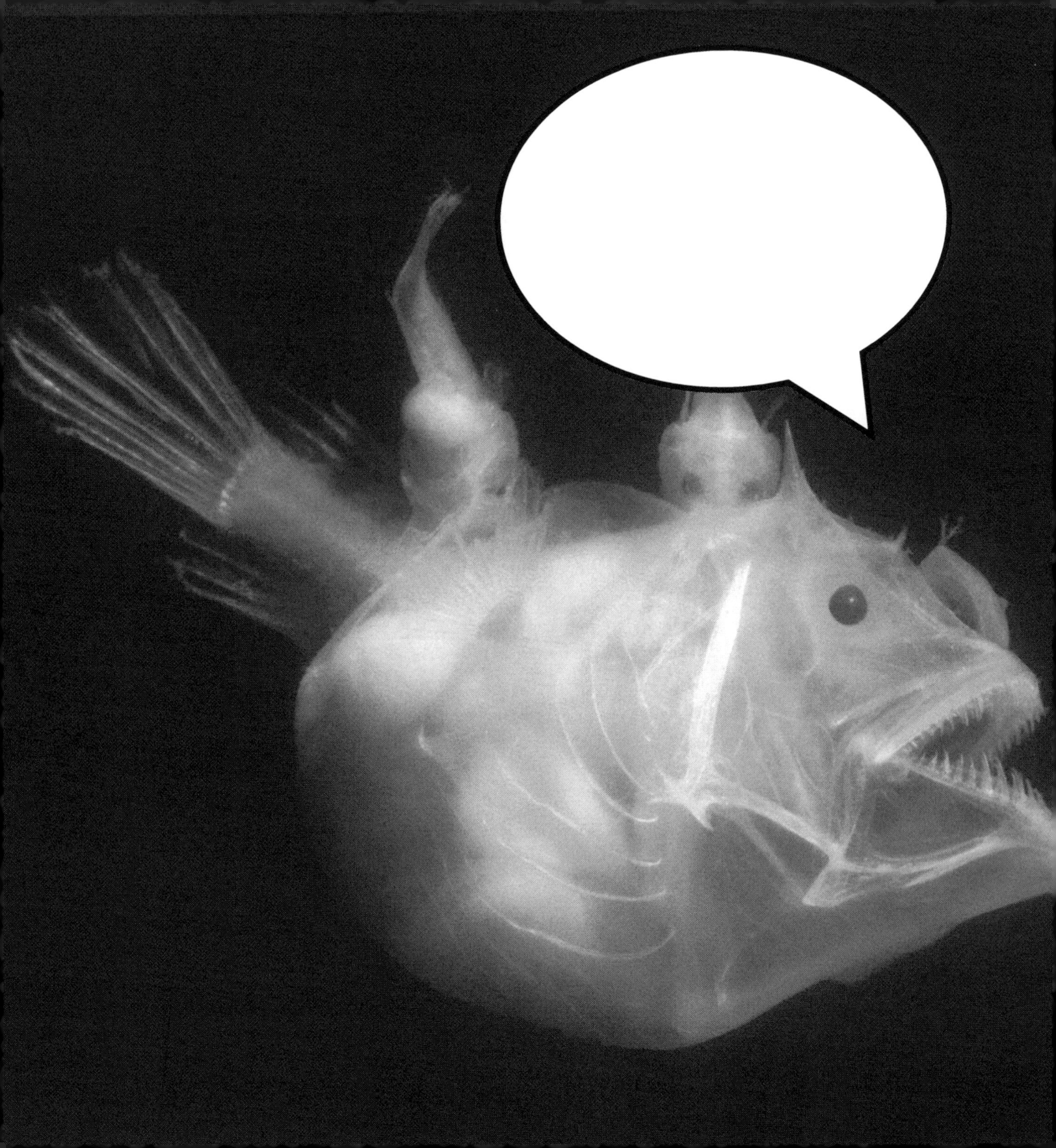

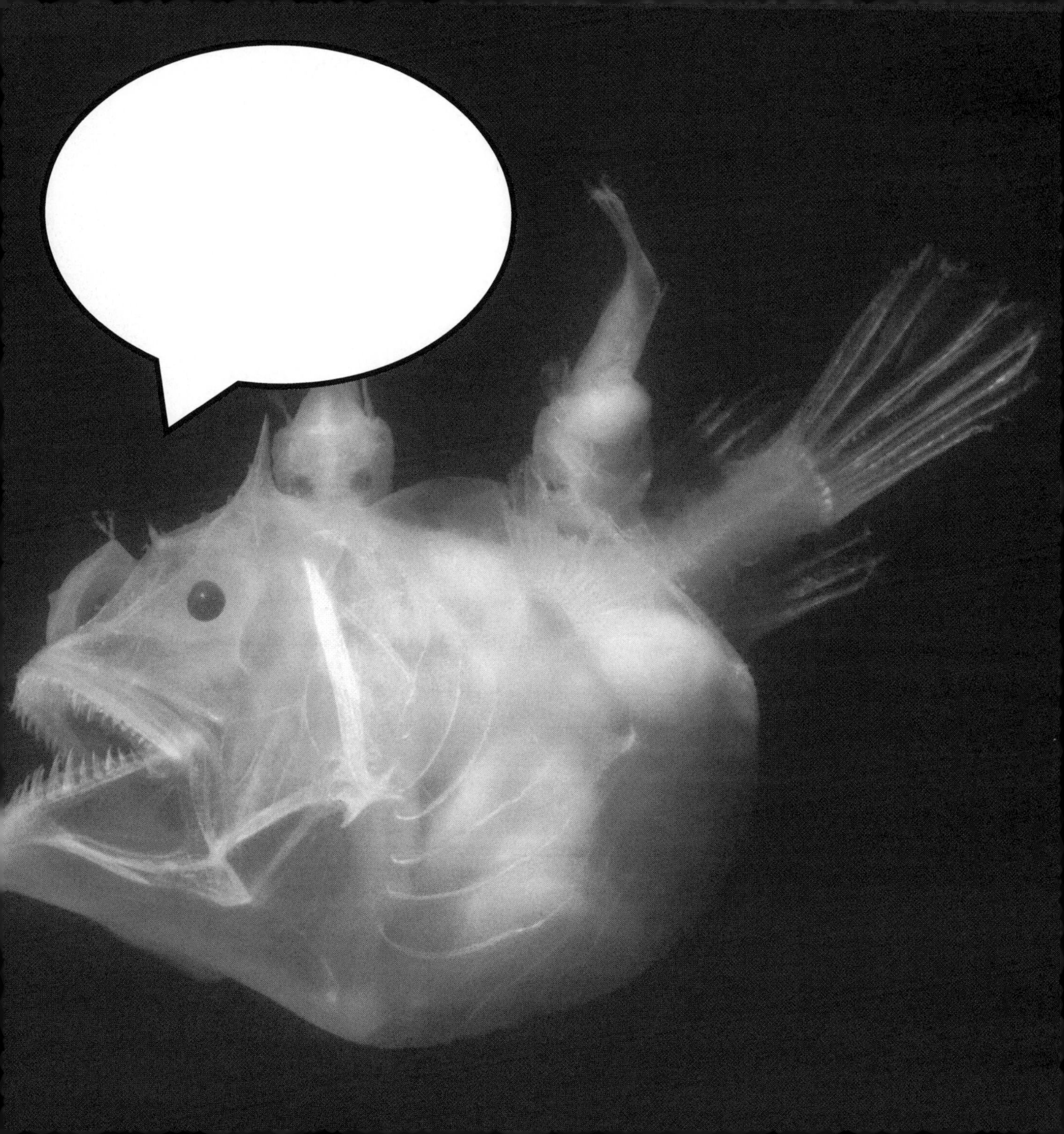

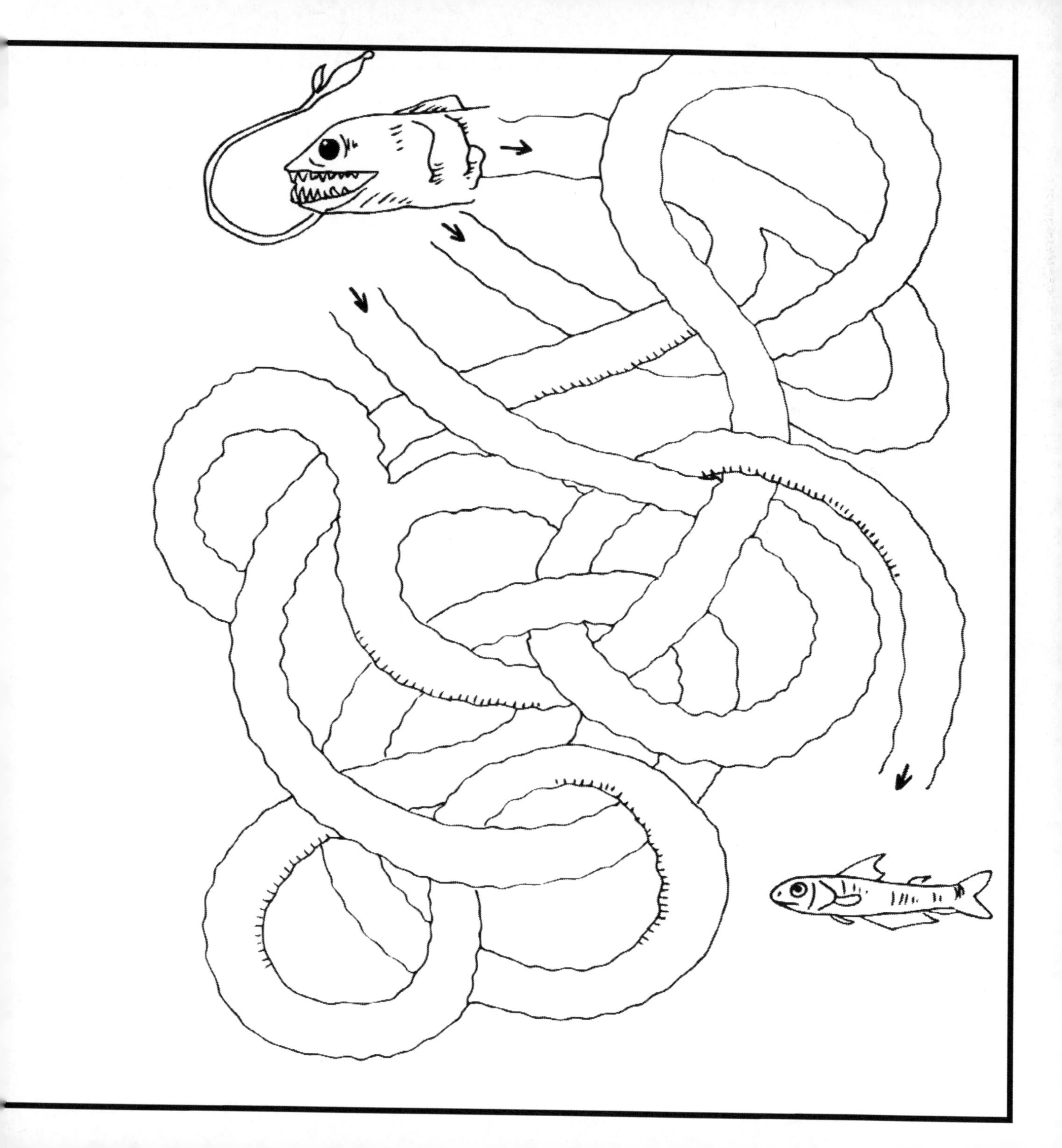

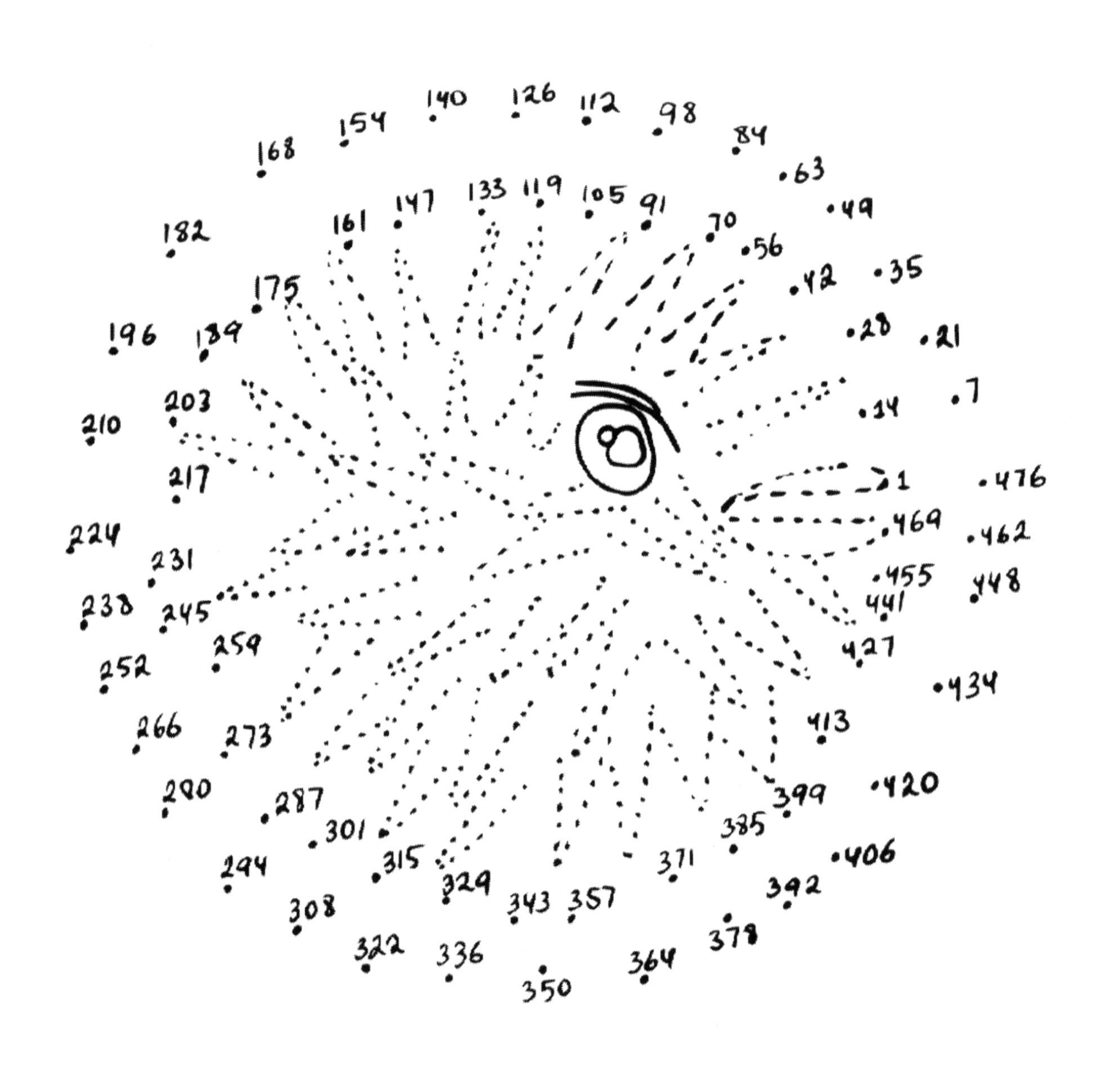
140
126
112
154
98
84
168
63
133
119
105
147
91
49
161
70
182
56
35
42
175
28
196
189
21
7
203
14
210
217
1
476
469
462
224
231
455
448
441
238
245
259
427
252
434
413
266
273
420
290
399
287
385
301
406
315
371
294
329
392
343
357
308
379
322
336
364
350

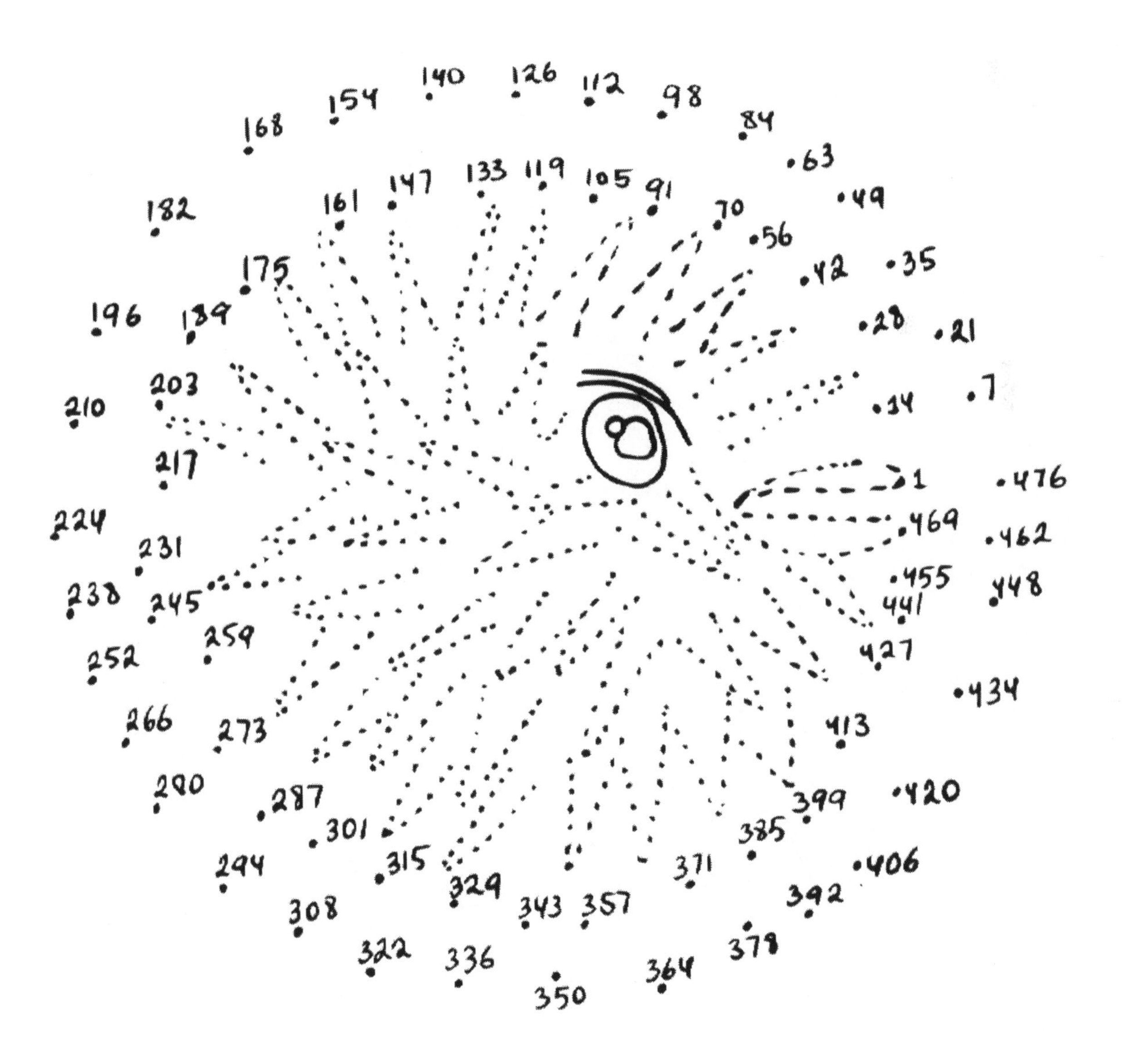

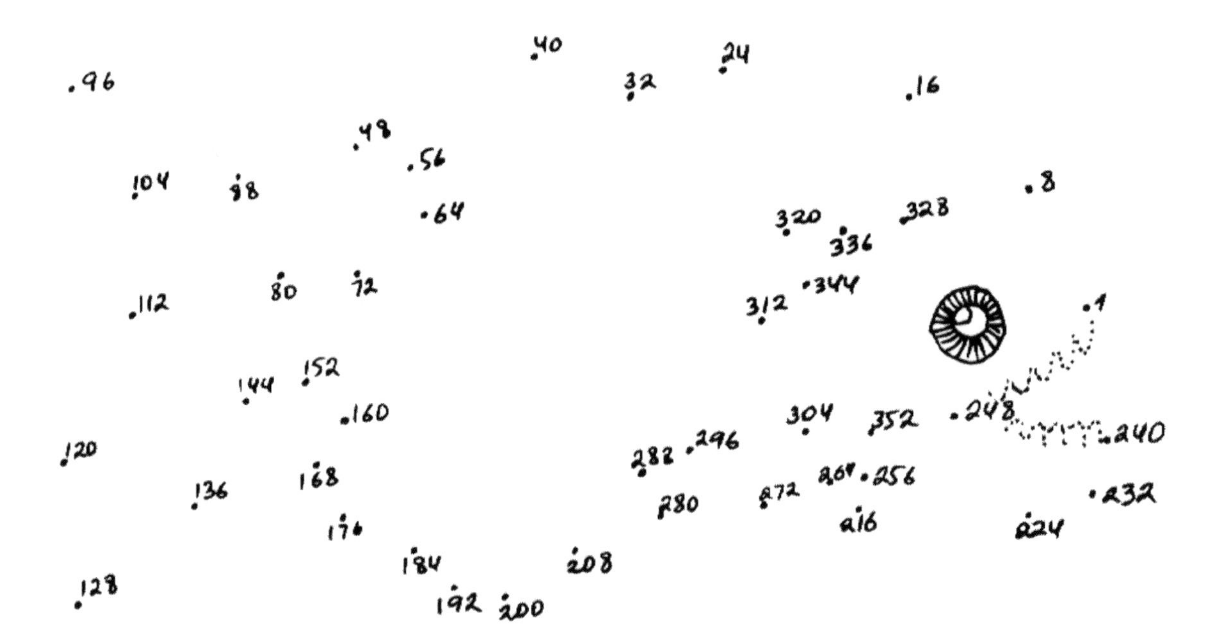

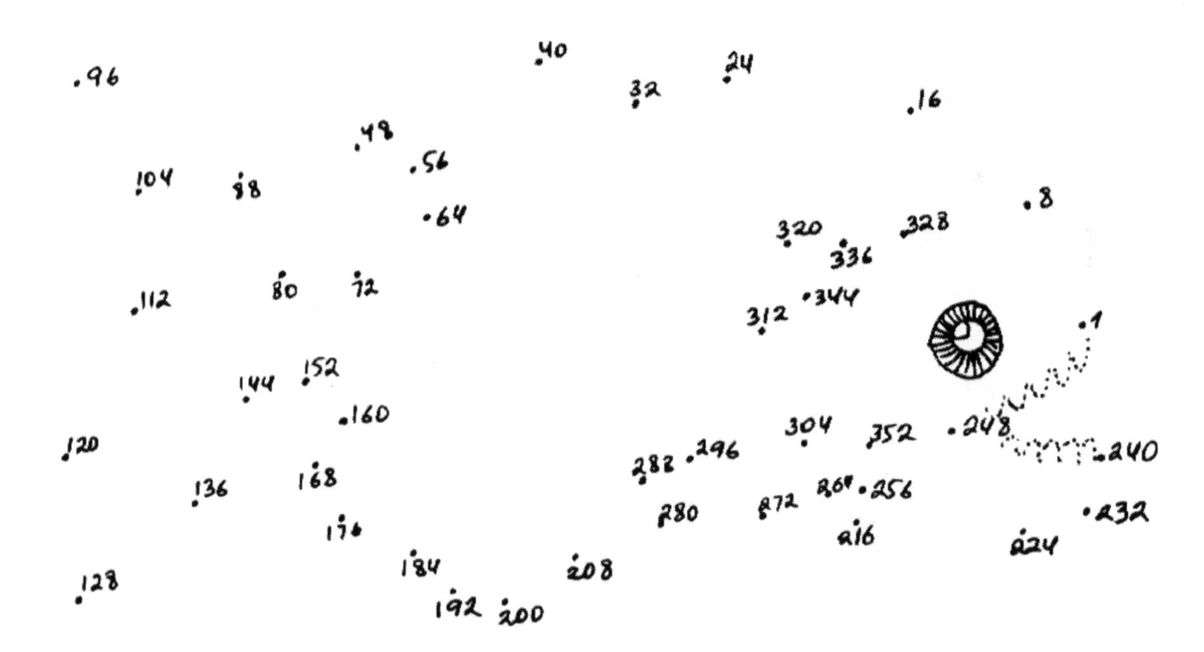

40
24
96
32
16
48
56
104
88
8
64
320
328
336
112
80
72
344
312
1
152
144
160
304
352
248
120
296
240
288
168
136
264
256
280
272
176
216
232
224
184
208
128
192
200

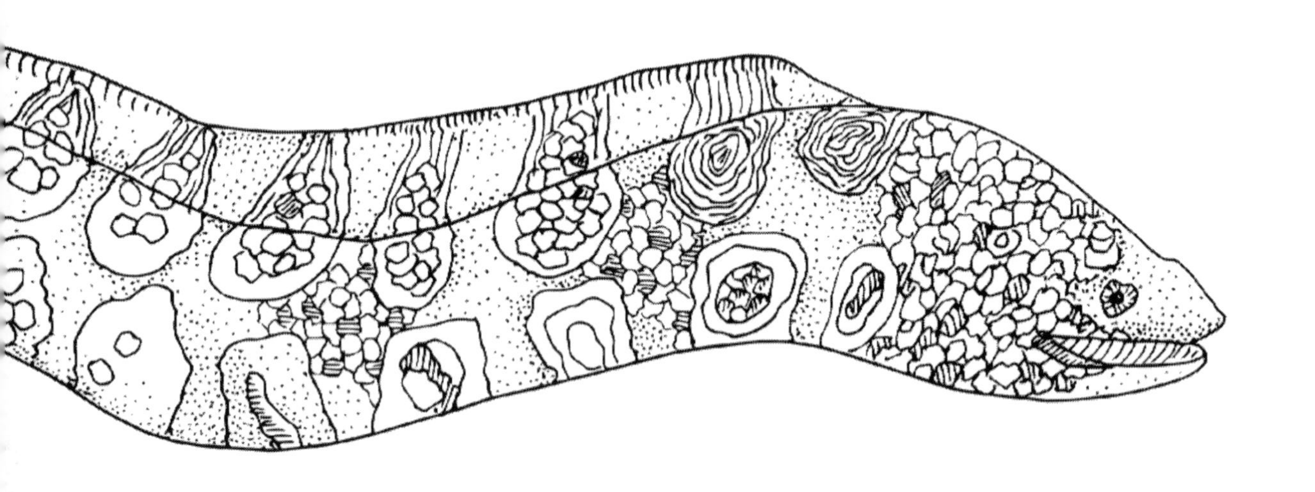

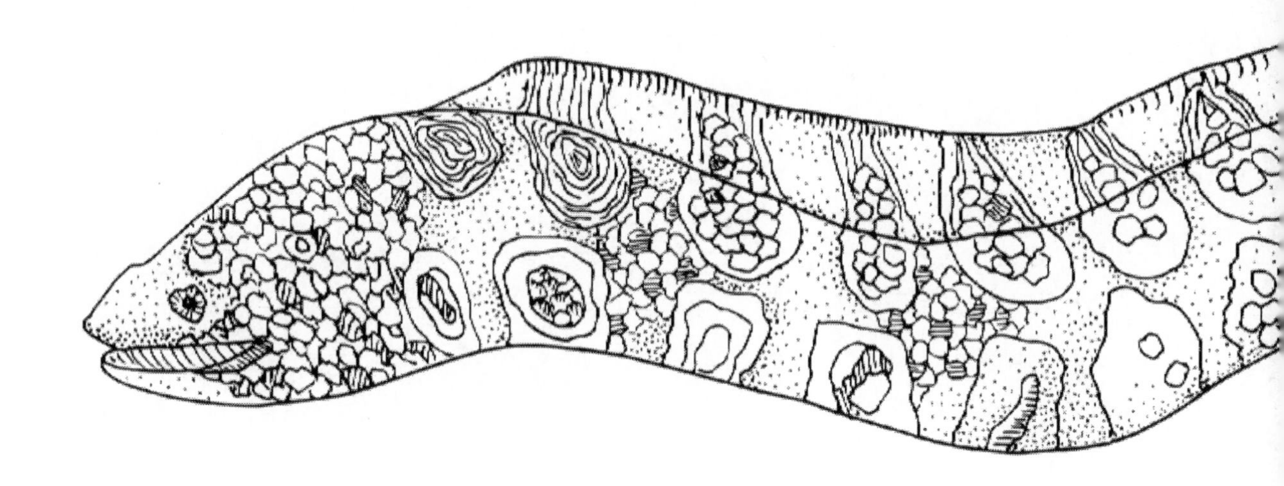

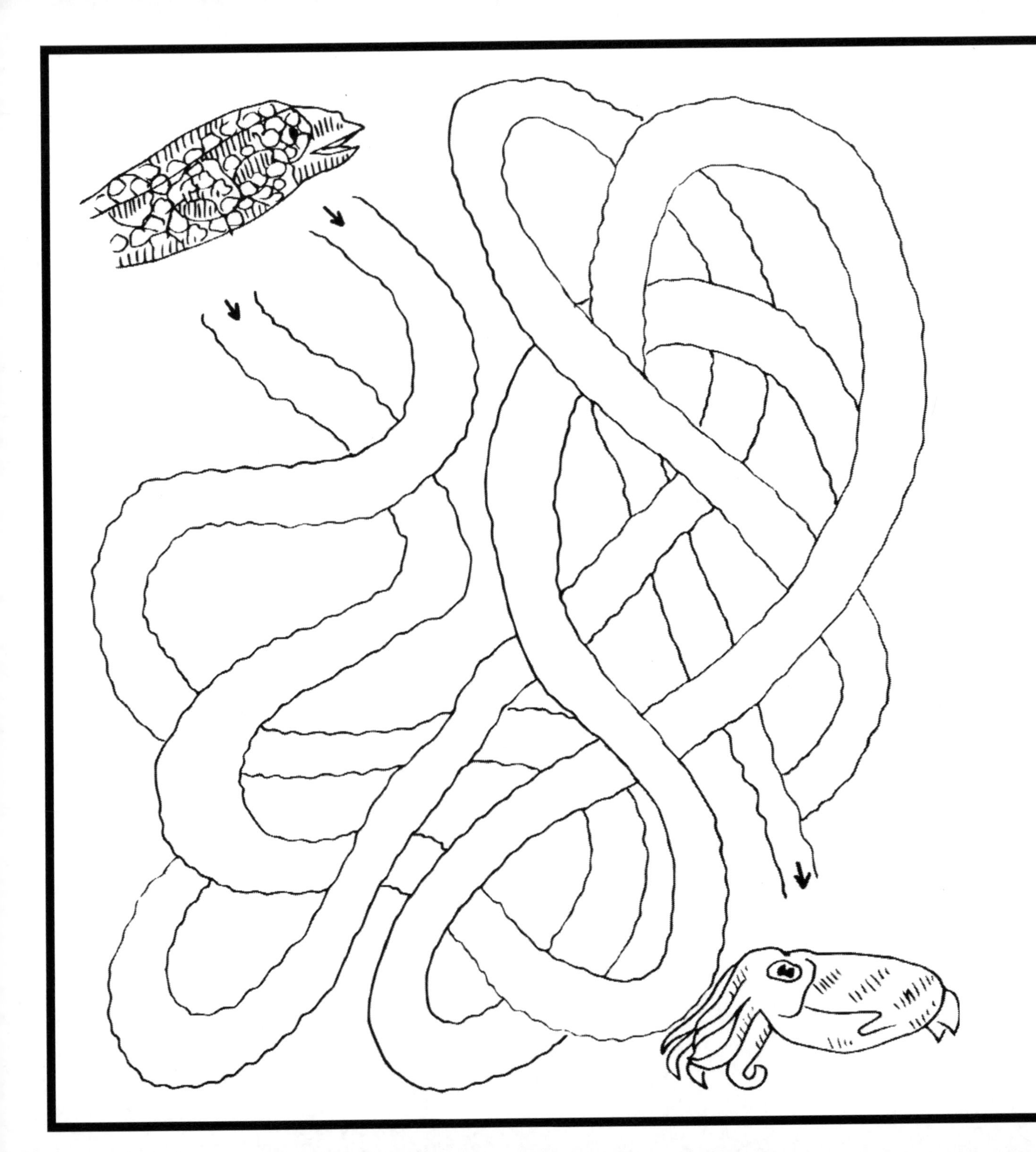

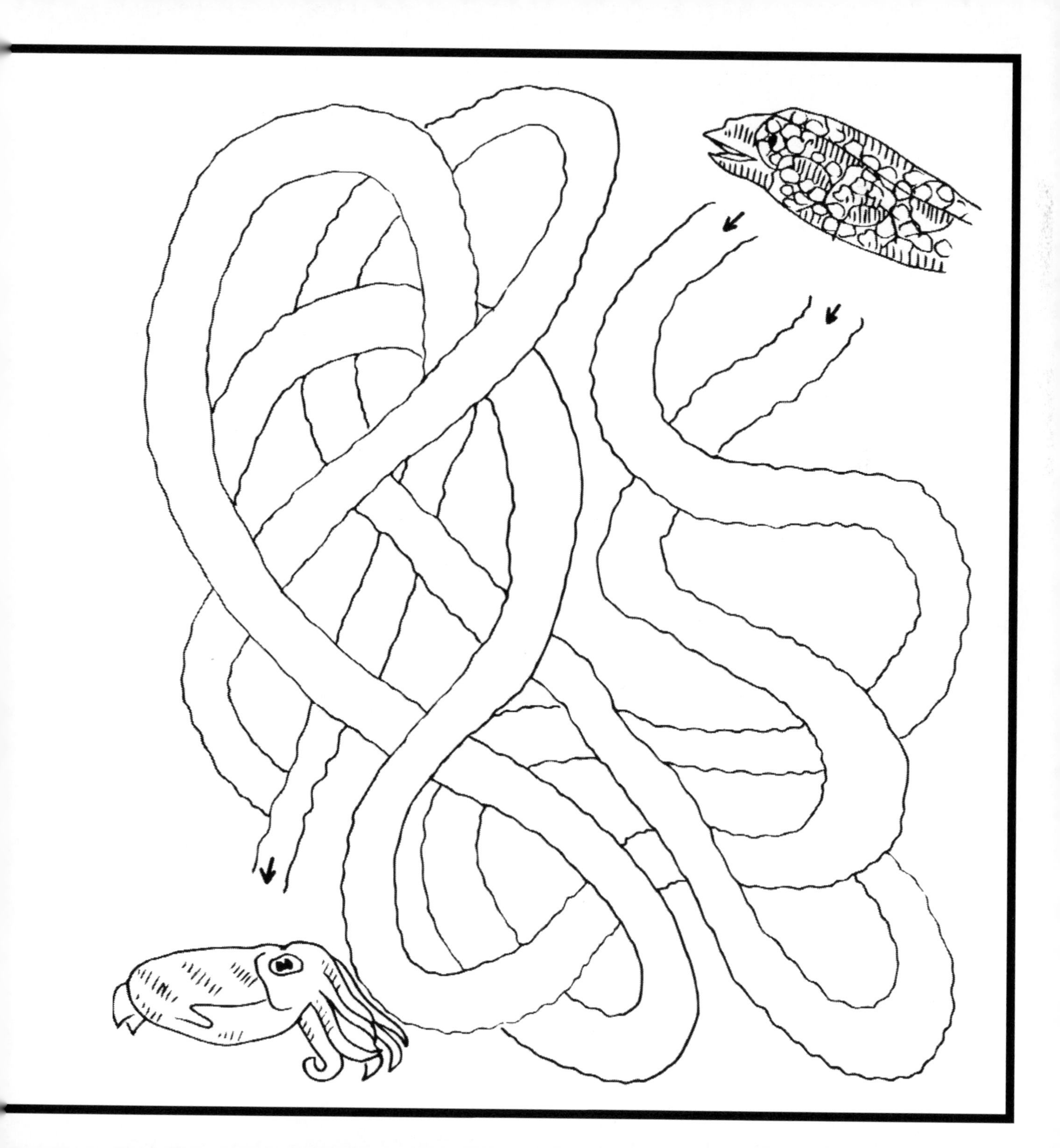

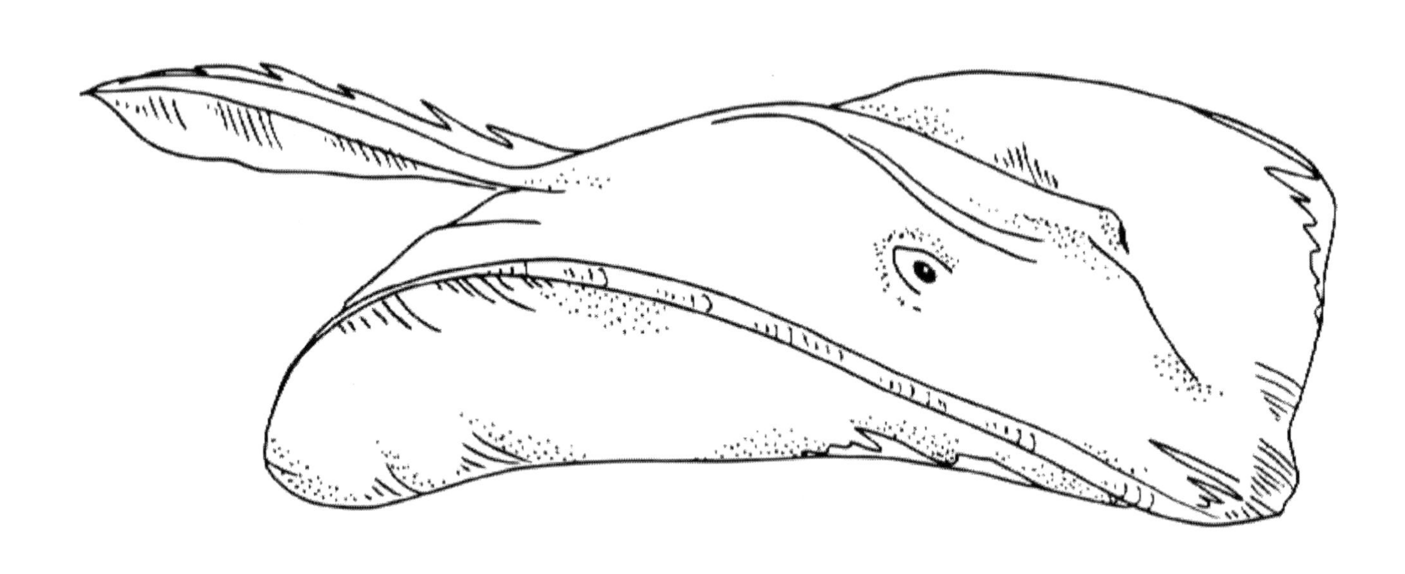

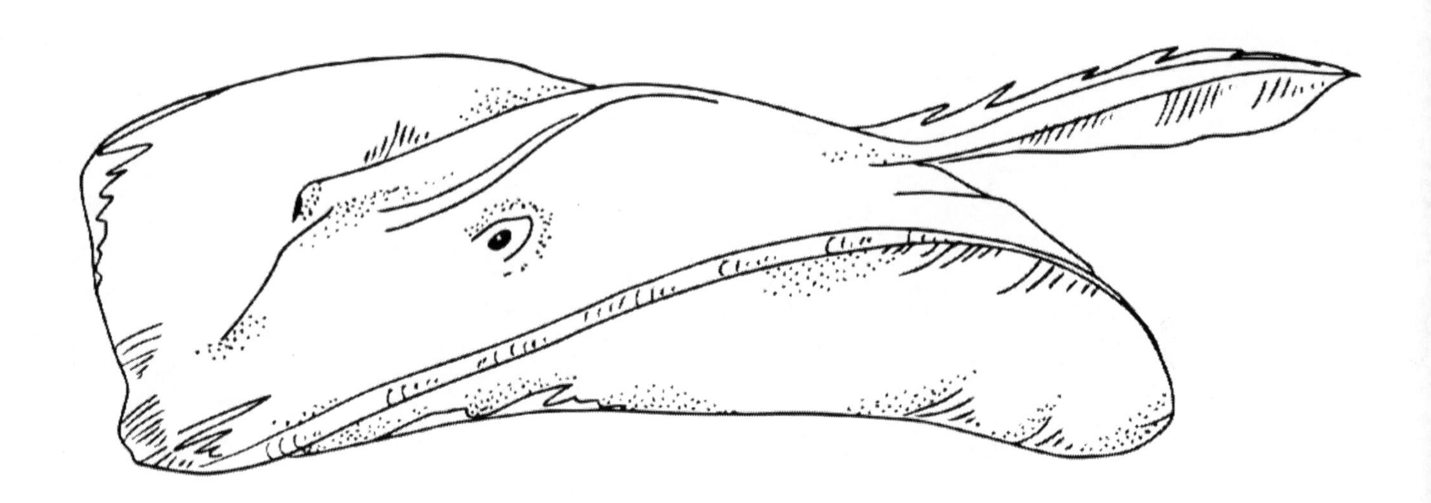

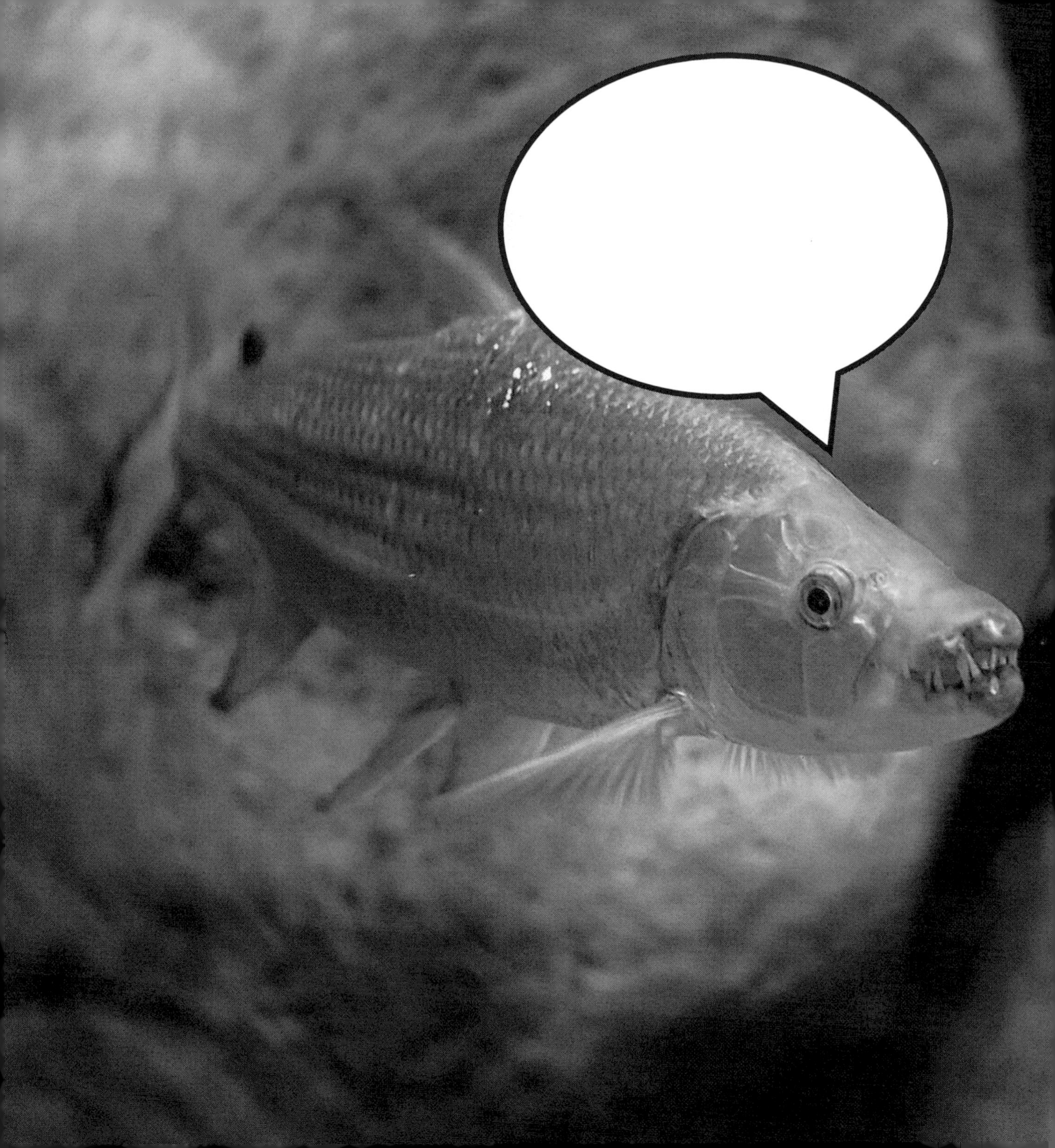

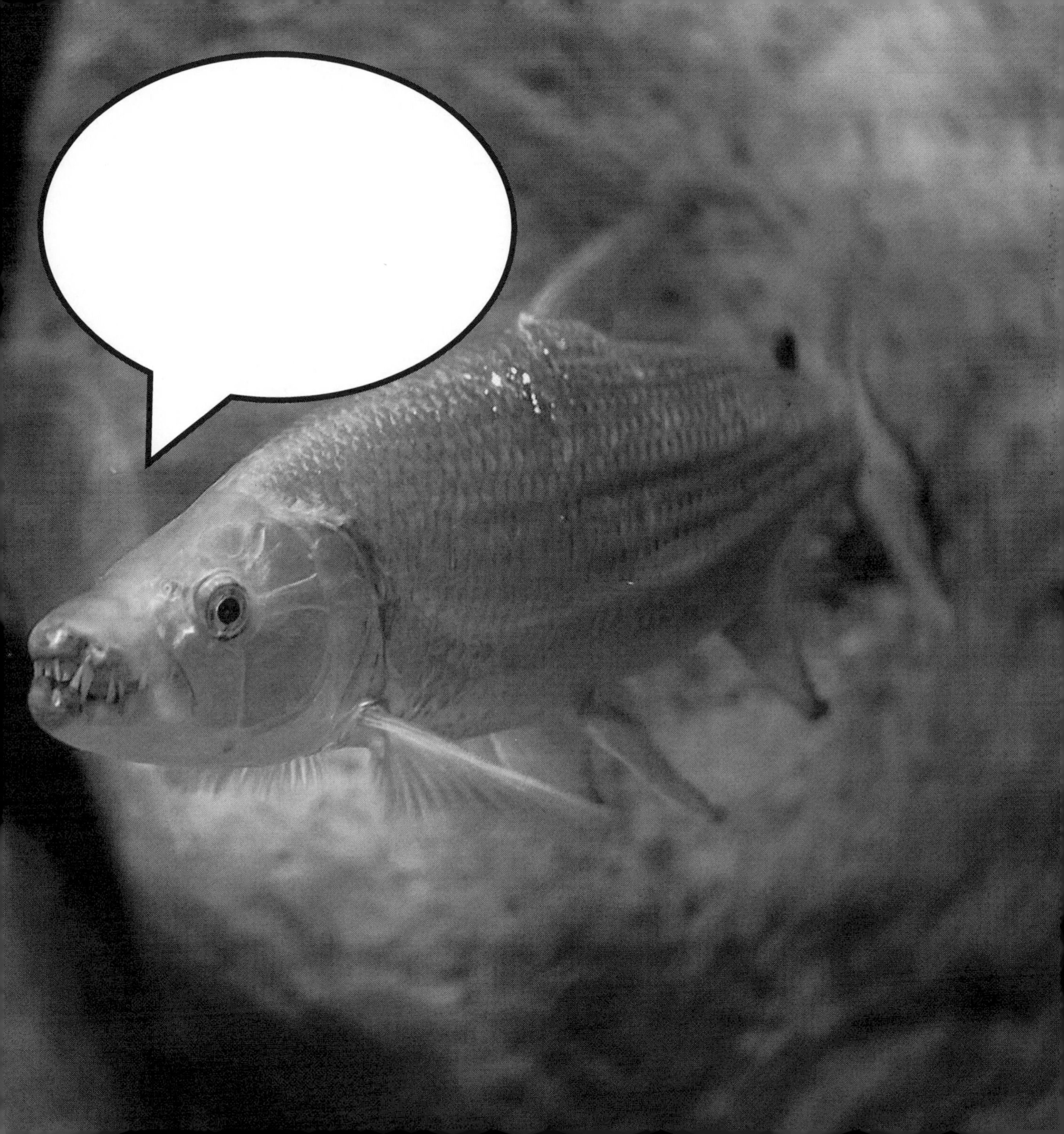

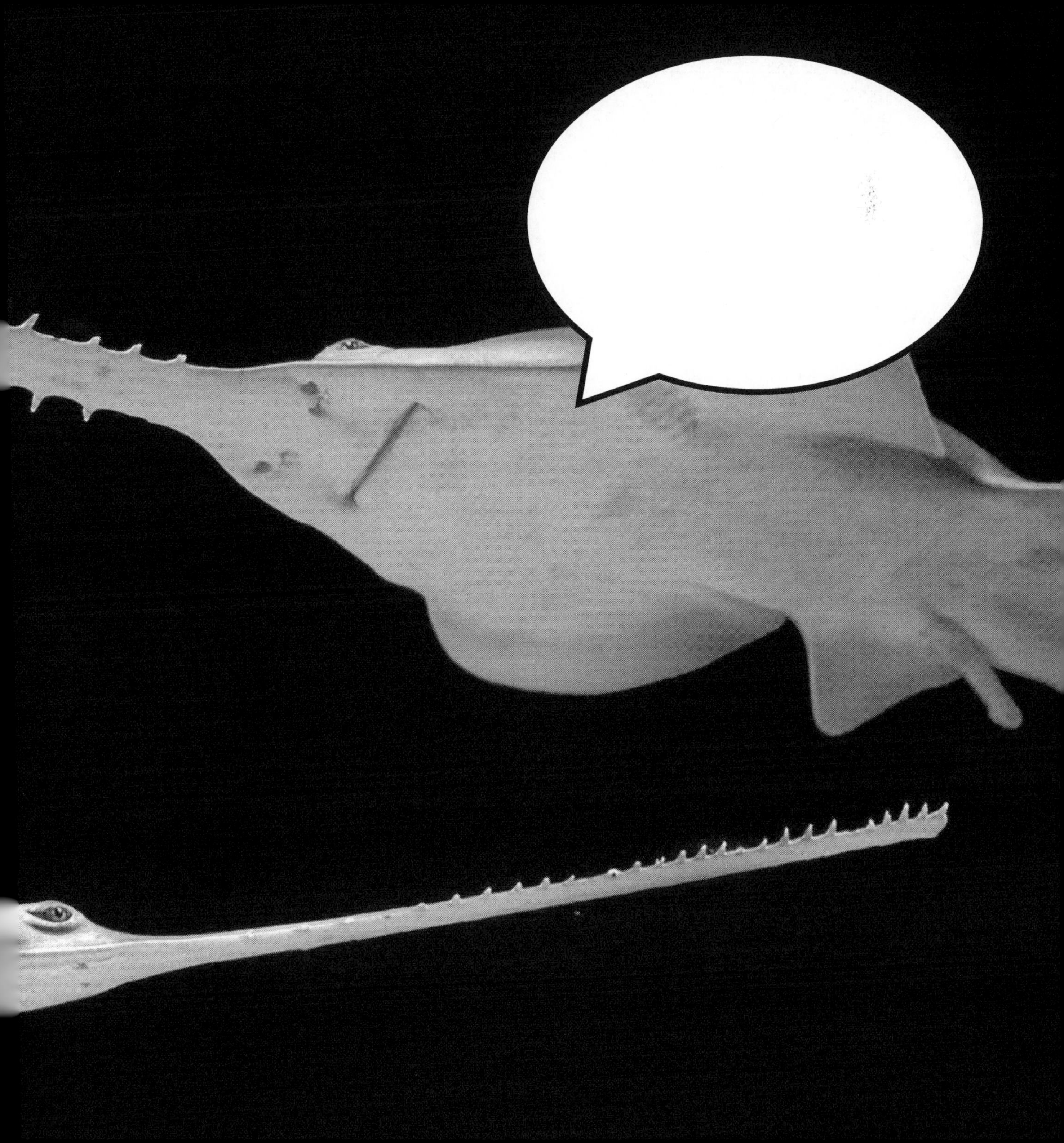

THE THINKING TREE

Made in the USA
Coppell, TX
05 February 2023